Urban Horticulture - Sustainable Gardening in Cities

Edited by Ali Kuden and Burhanettin İmrak

Published in London, United Kingdom

Urban Horticulture - Sustainable Gardening in Cities
http://dx.doi.org/10.5772/intechopen.105243
Edited by Ali Kuden and Burhanettin İmrak

Contributors
Lina M. Alnaddaf, Salim F. Bamsaoud, Mahroos Bahwirth, Mebarka Daoudi, Hadja Guedaouria, Youcef Benmoussa, Abdelhak Maazouzi, Karima Seddiki, Fouzia Mehdi, Hanane Maammeri, Asma Belaid, Tatenda Musasa, Kudakwashe C.R. Muringaniza, Emmanuel Dakarai Mhlanga, Arini Putri Hanifa, Eka Triana Yuniarsih, Retna Qomariah, Nurmalinda, Yopi Saleh, Yati Haryati, Indarti Puji Lestari, Susi Lesmayati, Kasey Stepansky, Theresa Delbert, Janet C. Bucey, Catherine Simpson, Cade Coldren, Ioana A. Coman, Caitlyn Cooper-Norris, Gad Perry, Mohd. Muzamil, Sehreen Rasool, Mohd. Mudasir Magray, Ummyiah H. Masoodi, Shabir Ahmed Bangroo, Ajaz Ahmed Malik, Saba Banday, Fernanda Da Conceição Moreira, Eglerson Duarte

First published in London, United Kingdom, 2023 by IntechOpen
IntechOpen is the global imprint of INTECHOPEN LIMITED, registered in England and Wales, registration number: 11086078, 5 Princes Gate Court, London, SW7 2QJ, United Kingdom

British Library Cataloguing-in-Publication Data
A catalogue record for this book is available from the British Library

Additional hard and PDF copies can be obtained from orders@intechopen.com

Urban Horticulture - Sustainable Gardening in Cities
Edited by Ali Kuden and Burhanettin İmrak
p. cm.
Print ISBN 978-1-83769-487-7
Online ISBN 978-1-83769-486-0
eBook (PDF) ISBN 978-1-83769-488-4

For EU product safety concerns:
IN TECH d.o.o., Prolaz Marije Krucifikse Kozulić 3, 51000 Rijeka, Croatia,
info@intechopen.com or visit our website at intechopen.com.

Meet the editors

Prof. Dr. Ali Küden is an academic staff member at the Çukurova University, Faculty of Agriculture, Department of Horticulture, and served as the Director of the Çukurova University Horticultural Research and Application Center between 1997-2017. Ali Küden is a full professor, and pomologist, studying on propagation and training of temperate fruit trees, especially pome and stone fruits, and nursery plants, at the University of Cukurova. He has been also working for 37 years on the selection and collection of fig and almond germplasm at the University of Cukurova. He has been extending his theoretical and practical knowledge to the growers, nurseries, and private sector in the whole country. He has published more than 170 scientific papers. He is one of the members of the Temperate Fruits in the Tropics and Subtropics (TFTS) and the Working Group of the International Society for Horticultural Science (ISHS). In addition, he took on the role of the organization of some scientific meetings of several organizations in and out of the country. He has been involved in several national and EU Prima projects, Prof. Dr. Ali Küden is married and has two children.

Assoc. Prof. Burhanettin İmrak is an academic staff member at the Çukurova University, Faculty of Agriculture, Department of Horticulture, His research interests include sustainable farming, fruit growing, physiological studies, and fruit breeding, physiology of dormancy in temperate climate fruits, and the effects of climate change on fruit growing. He has made short- and long-term visits and training studies in many countries on these issues. Between 1999 and 2000, he attended a plant breeding course organized by CIHEM in Zaragoza, Spain. Dr. İmrak has written books and book chapters and published around 98 research articles. He has served as a referee for journals. He has participated in many international symposiums, several workshops, and national meetings. He has been working on EU Prima and Cost projects. Dr. İmrak is married and has two children.

Contents

Preface

The Edited Volume *Urban Horticulture – Sustainable Gardening in Cities* is a collection of reviewed and relevant research chapters, concerning the developments within the book's field of study. The book includes scholarly contributions by various authors and is edited by a group of experts pertinent to the horticulture field. Each contribution comes as a separate chapter complete in itself but directly related to the book's topics and objectives.

The book includes chapters dealing with the topics: "Perspective Chapter: Application of Nanotechnology Solutions in Plants Fertilization and Environmental Remediation", "Hydraulic Sizing for Watering Green Space Application in Bechar-Algeria", "Perspective Chapter: The Impact of Human Activities on Wetlands' Provisioning and Cultural Services in Epworth, Zimbabwe", "Perspective Chapter: How Important is Urban Farming in Indonesia to Support Food Sovereignty?", "Therapeutic Impact of Engagement in Green Spaces", "Urban Vegetation: Anthropogenic Influences, Public Perceptions, and Wildlife Implications", "Perspective Chapter: Mechanization in Agricultural Production from Horizontal and Vertical Perspective", "Landscaping Promoting Sustainable Comfort in Cities". The target audience comprises scholars and specialists in the field.

IntechOpen

Chapter 1

Perspective Chapter: Application of Nanotechnology Solutions in Plants Fertilization and Environmental Remediation

Lina M. Alnaddaf, Salim F. Bamsaoud and Mahroos Bahwirth

Abstract

The effects of nanoparticles that are used on plants, either as foliar sprays or as fertilizers, vary between promoting and inhibiting. This effect varies according to many different factors, such as the type of nanoparticles, the concentration, the shape, the size, the type of plant, the soil characteristics, and the soil microorganisms. The effect of iron, zinc oxide, graphene, copper oxide, silicon, titanium, and carbon nanotubes on soil fertility, plant growth and development, and crop yield was discussed in detail. The nanoparticles affect the seed's water absorption, roots, germination, stem, photosynthesis rate, photosynthetic pigments, and enzymatic and non-enzymatic compounds. Moreover, it also highlights the role of these particles in the different stresses that can be exposed to the plant and the mechanisms of tolerance of these stresses. This chapter presents the ability of these particles to combat pollution in its various forms, including groundwater, heavy metals, and wastewater. In addition, these nanoparticles accumulate in the water, soil, and plants, and impact humans and the food chain. Finally, the future prospects for the use of nanotechnology to achieve the goals of sustainable development.

Keywords: nanotechnology, fertilization, environmental remediation, nanoparticles, plant

1. Introduction

Nanotechnology can change agricultural systems and methods used *via* nanofertilization. It contributes to reducing pollution and protecting the environment, as well as reducing the costs of using chemical fertilizers. Nanofertilizers provide soluble nutrients that are quickly absorbed and reduce their build-up in the soil. In addition, it increases plant growth and development and enhances plant resistance to biotic and abiotic stresses. Nanofertilizers' effect on plants varies according to the plant type, method of application, and the size and concentration of different nanoparticles. Moreover, to the positive effects of nanofertilizers, it is necessary to highlight their adverse effects related to toxicity. In addition, we will discuss the future prospects for the use of nanofertilizers.

2. Mechanisms of nanofertilizer impacts on soil, plants, crops, and stress

2.1 Soil fertility

A fertile soil is one of the fundamentals (along with water, climate, and seeds) that are essential for the advancement of agriculture and enhancing agricultural output efficiency. Several groups of researchers each suggested a creative approach to soil remediation. Advances in nanotechnology over the last century have allowed for a variety of approaches to increasing soil fertility. Unpredictably, NPs in soil may either inhibit plant development or promote it, depending on many factors related to nanoparticles and plants [1]. Several techniques, including spraying, pouring down nanofluid, and combining nanoparticles with the soil, are used to amend the soil with nanoparticles in order to increase soil fertility. Brownian motion and gravitational force may both play a role in the transit of NPs in soil, hence increasing the likelihood of these particles interacting with soil constituents that subsequently reflected plants [2]. If NPs are tiny enough (5 to 20 nm), they may penetrate plant cells through sieve-like cell walls [3]. This has enhanced the need for employing the substance in nanoform to overcome the problems in soil fertility and plant fertilization.

Several soil-beneficial processes are attributed to the use of nanomaterials, including the nitrogen cycle, the enhancement of enzyme activity, and the induction of plant-beneficial microbes. Numerous inorganic, organic, and composite nanomaterials were examined in terms of the possible influence they might have on fertilizing soil and resolving issues connected to plant development and production. These studies were conducted both in the laboratory and in the field, and they were evaluated and summarized in order to be a useful resource for farmers and researchers to save time and money.

One of the ways that nanoparticles are employed in agriculture is by incorporating them (in either fluid or solid form) into the soil [4]. These nanoparticles have the potential to introduce either on their own or in combination with conventional fertilization ingredients. In each of the aforementioned methods, nanoparticles and nanotechnology, in general, have the potential to provide solutions to soil problems that have preoccupied farmers over the period of the last several years. Among these solutions was the application of silver nanoparticles on the soil. The combination of these nanoparticles caused a rise in the number of diazotrophic bacteria in the soil, which is thought to have been a contributing factor in the elevated amount of fertilization seen in the soil. Moreover, iron nanoparticles, which are a component of a new generation of environmental remediation technology, may provide a cost-effective solution to one of the most challenging difficulties in environmental cleanup. Soil contaminants such as chlorinated organic compounds, toxic metals, and inorganic compounds can all be converted into less toxic or inert compounds by interacting with nanoparticles zero-valent iron [5–9]. On the other hand, it is shown that soil amendments containing metallic Cu NPs up to 600 mg/kg considerably improved the development of lettuce seedlings by up to 91% without causing any harmful consequences [1]. Recent research has shown that adding zinc nanoparticles to the soil as a kind of fertilizer considerably increases the amount of zinc that is available in the soil, which in turn significantly increases the amount of zinc that is present in the plant's shoots and grain. This is due to that the application of these fertilizers considerably accelerated the development of plant roots and exudates. These exudates have the potential to greatly boost both the availability of zinc and its absorption by plants [10, 11].

One of the most promising approaches that deliver the answer to reducing soil toxicity is the use of nanocomposites, which will eventually result in an improvement in plant growth and production [12]. The higher surface-to-volume ratio of graphene oxide (GO) nanoparticles has unique capabilities as an efficient water transporter in soil because of its sp2 and sp3 hybrid structure; these particles could contribute to solving drought problems and the lack of water [13]. CuO nanoparticles, on the other hand, promote the growth of plant growth-promoting microorganisms in the rhizosphere of Salvia miltiorrhiza (red sage) [14, 15]. The ameliorating potential of Si nanoparticles for soil is well known since these nanoparticles are often more effective than their bulk equivalents [12, 16–21]. Due to the stimulation of plant photosynthesis by SiO_2 nanoparticles, silicon plays a critical role in improving the plant antioxidant system's efficiency, promoting plant growth under salt stress, and increasing the plant's tolerance to abiotic and biotic stressors. Therefore, SiO_2 nanoparticles provide excellent solutions for enhancing the capabilities of the soil to support agricultural activities [12, 22]. TiO_2 nanoparticles have several uses in agroecosystems due to their attractive physical and chemical characteristics, availability, cheap cost, and excellent stability. TiO_2 NPs are used in nanofertilizers and nanopesticides to increase soil fertility and crop development as well as improved light retention [23–25]. Numerous studies show that TiO_2 nanoparticles, at a specific concentration range in various plant species, may increase plants' resistance to cold, heat, drought, NaCl, and ultraviolet light [26–31]. TiO_2 NPs stimulate soil nutrients and microbial dynamics at low concentrations [32]. Nanoparticles of titanium dioxide make it easier for seeds to absorb water and, as a result, accelerate the germination process [12]. The radicle, plumule, root, and seed germination of canola seedlings is accelerated by the application of TiO_2 NPs [33].

Carbon nanotubes are employed as fertilizer to enrich the soil, consequently encouraging the growth of many plants, including tomatoes, cucumbers, rice, maize, broccoli, and soybeans [3, 34–39]. Carbon nanoparticles have the ability to restrict the mobility of herbicides in agricultural applications. Furthermore, CNPs improve plant photosynthesis, crop development, water absorption, antioxidant levels, and the efficiency of nitrogen (N), P, and K use. CNPs have recently reported as part of a novel approach that also includes compost, mycorrhizal, and arbuscular fungi blended into the soil. This combination of CNPs has the ability to protect plants against drought stress. The innovative use of CNPs in agriculture might fulfill growing food demand while simultaneously protecting the environment [40]. The recent approach is a significant indicator that nanotechnology is a fertile field for agricultural research, and many more studies will be required to develop innovative agricultural solutions in the near future.

2.2 Plant growth and development

The use of nanotechnology in agricultural sciences, with the assistance of nanoscience, is now achieving a high level of success, resulting in the development of novel solutions in a wide range of agricultural domains. The consequences of all the novel approaches to enhance soil, irrigation water, and agricultural climate discovered *via* the usage of nanotechnology are indirectly improving plant components in order to produce agricultural items and boost yields. For example, the addition of iron nanoparticles to the soil reduces soil contaminants such as chlorinated organic compounds, toxic metals, and inorganic compounds that could harm plants. A wide variety of nanoscale materials were prepared and analyzed to provide a fresh strategy for addressing agriculture's perennial challenges. These nanoparticles are often delivered directly to plants through the seeds, stems, or leaves in order to improve growth and agricultural productivity. In this section, all of the recent researches on how nanoparticles affect plant development are summarized.

Using carbon-based nanomaterials (CNTs) boosted seed water absorption by moving it from the seed coat to shoots and leaves [1]. This improved germination percentage and plant growth by maximizing water uptake and nutrient intake [41]. This technique has been used to improve seed germination, root and plant growth, and final growth in numerous crops, including hybrid Bt cotton, *Phaseolus mungo* L., *Brassica juncea* L., tomato (*Lycopersicon esculentum* Mill) and rice (*Oryza sativa* L.), resulting in an increase in biomass and seed germination [12]. CNTs have a crucial role in stimulating rice seedling growth, root elongation, and seed germination in zucchini species [42]. In addition, CNTs improved seed germination, vegetative biomass, and tomato growth [43], and multi-walled carbon nanotubes (MWCNT) influenced the fresh and dried mass of tomato roots as well as gene expression [44]. Research demonstrates that maize, tobacco, switchgrass, rice, tomato cell cultures, barley, wheat, and soybean seeds grew faster when treated with single-walled carbon nanohorns (SWCNHs) [12]. In fact, the effect on plant shape depends on the type of nanoparticles and its applied [45]. However, titanium dioxides (TiO_2) NPs, such as Ag and graphene NPs, have shown to improve seed germination by increasing the ability of internal tissues to convey water and increasing the metabolism of seed stores [24]. They also improve water absorption, which accelerates seed germination [12]. Increased nitrate absorption and photosynthetic rate attributed to the augmentation of nitrogen metabolism caused by the introduction of TiO_2 NPs in spinach, showing that these nanoparticles may really boost plant growth [44]. The application of TiO_2 NPs to canola seedling plants increased the growth of the radicle and plumule, as well as root and seed germination [33].

Nanoscale zinc oxide (ZnO) is essential for plant growth at low concentrations and does not inhibit onion seedling growth or cell division [33]. ZnO NPs enhanced the germination of soybean, wheat, tomato, and onion seeds [46] and increased the germination of cucumber seeds by 10%. In addition, ZnO NPs with a particle size of 50 nm favorably influenced the roots of rapeseed [33]. In addition, ZnO NPs application on the coffee plants had a positive effect by increasing the fresh weight of roots (37%) [44]. Compared to the control, ZnO NPs at 10 mg/L accelerated seed germination (100%), root length (185 mm), and root width (0.5 mm) [47].

Similar to other nanomaterials, silver nanoparticles (SNPs) had beneficial impacts on vascular plants, including seed germination, root development, and plant biomass. These benefits correlated with the concentration and structure of SNPs [42]. Experiments conducted with varied concentrations of SNPs on Cucurbita Pepo revealed a notable change in seeding speed and duration due to the presence of the SNPs [48]. SNPs greatly improved wheat seeding rate, rootletlet and plumule length, and rootletlet and plumule wet and dry weight [49]. In fenugreek plants, low concentrations of SNPs (10–20 g/mL) increased seedling growth and seed germination [44]. A number of investigations on the impact of SNPs on two wheat and barley types found a rise in germination ratio stem, length, and a decrease in root length compared to the control [50].

Rice seed germination is stimulated by the addition of Silicon (Si) NPs. Si NPs increase maize seed germination by improving nutrient availability [41]. Si NPs enhanced all seedling parameters such as percent germination, germination rate, length, and fresh and dry mass of root and shoot in Changbai larch (*Larix olgensis* Henry) seedlings [51, 52]. Also, utilizing Si NPs for pre-chilling seeds in tall wheatgrass (*Thinopyrum intermedium* L.) breaks inertia, stimulates seed germination, and boosts the vigor index, mean germination time, and dry weight of seedling roots and shoots [50]. Soybean (*Glycine max* L.) germination and growth are boosted by boosting nitrate reductase activity and improving the seeds' capacity to absorb and use water and nutrients [51, 52]. Seed germination, plant height, and root and shoot dry weights were all improved by using Si-NPs in rice (*Oryza sativa* L.) seedlings [53]. In a similar manner, Se NPs stimulated organogenesis and accelerated root development in *Nicotiana tabacum* L. by up to 40%, compared to the effect of aqueous selenate [33]. Gold nanoparticles (NPs) at concentrations of 5–15 mg/L considerably enhanced the germination and physiology of older maize seeds without causing any harm [54].

Numerous researchers have used a wide number of techniques, including the biological approach, in order to fabricate NPs with a wide range of morphologies and sizes [48, 51, 52, 54–57]. The biological approach is a synthetic technology that involves the extraction of plant components for use in the nanoparticle synthesis process. This technology has the potential to be an effective solution in agricultural applications since it is controllable, inexpensive, low-risk, and safe. Generally, nanotechnology has the potential to improve agriculture's sustainability and benefits, but there are still many questions surrounding its use in this field.

2.3 The effect of nanofertilizer on crop yield

Numerous studies outline the nanotechnology mechanism for crop enhancement. In this part, these findings on the function of nanotechnology in crop improvement are discussed. Nanotechnologies have the potential to generate a substantial increase in agricultural productivity as well as enhancements to food production systems [58]. Nanotechnology is one example of a technology that may aid in agricultural production by reducing wasteful input consumption and increasing productivity. Nanotechnology is regarded as an effective tool for managing a wide range of environmental stresses by providing novel and practical solutions. Nano-fertilizers (NFs) considerably increase plant growth efficiency, soil quality, and crop yield of high-quality fruits and cereals [59], and also, may serve as a sustainable strategy to promote agricultural yield in (semi-) arid regions [60]. NFs have the potential to improve nutrient absorption and plant output by modulating fertilizer availability in the rhizosphere, prolong stress tolerance by enhancing nutritional capacity, and raise plant defense mechanisms. For sustainable agriculture, NFs might potentially take the place of synthetic fertilizers since they are better at stimulating plant growth [61]. The applications of nanoparticles and nanomaterials have a positive effect on crop productivity *via* different strategies. It is summarized by green synthesis of nanoparticles, plant-targeted protection *via* the application of nanoherbicides and nanofungicides, precise and constant supply of nutrients through nanofertilizers as well as tolerance to abiotic stress, by several mechanisms such as activation of the antioxidant enzyme system that alleviates oxidative stress [62]. In addition, the tolerance to abiotic stress by several mechanisms, such as activation of the antioxidant enzyme system that alleviates oxidative stress is one of the positive impacts of the applications of nanoparticles and nanomaterials in agriculture. All the mentioned positive effects could be due to their nano-size properties, their high nutrient use efficiency, their slow release of nutrients, and the controllable necessary dosage of nanomaterials used in fertilizer to get the desired results [63].

In the most recent decades, there have been reports that nanoscale zero-valent iron has the ability to increase plant development in the laboratory. Therefore, nanoprimed seeds demonstrated superior crop performance when compared to hydroprimed seeds in the conventional sense. As a result, nZVI is classified as a "pro-fertilizer," and it is utilized commercially as a seed treatment agent capable of increasing plant development and yield while causing little disruption to the soil ecology [64].

Essential oil content was found to increase with the use of nano-iron fertilizer, fruit output and quality were both enhanced with the use of nano-zinc and boron fertilizers, and the use of nano-zeolite was shown to increase germination rates and prolong the life of soil nutrients. Increased growth from nanoparticles helps crops mature sooner and recover more quickly from environmental challenges. In addition to increasing the levels of photosynthetic pigments and osmolytes, the synthesized Fe_2O_3 NPs also increased the activities of the enzymes peroxidase (POD), polyphenol oxidase (PPO), catalase (CAT), and superoxide dismutase (SOD) in healthy as well as infected tomato plants when compared with the control. The optimum therapy for peroxidase and polyphenol oxidase activities was determined to be the administration of Fe_2O_3 NPs (10 g/mL) to stress plants, which enhanced the activities of POD by 34.4% and PPO by 31.24%. On the other hand, the best treatment for stressed plants was to apply Fe_2O_3 NPs (20 g/mL), which increased the activities of CAT by 30.9% and SOD by 31.33%) [65]. When compared to the control, the same treatment had significantly higher biological yield, grain yield, protein yield, and harvest index, with 14.792 Meg/ha, 7.100 Meg/ha, 890.34 kg/ha, and 40.00 percent, respectively [66]. Under normal circumstances, applying ZnO NPs led to considerable growth and biomass augmentation while preventing drought-induced decrease. ZnO NPs treatment increased photosynthetic pigments, photosynthesis, and PSII activity, with peak values at 100 mg/L [67]. Under both normal and drought circumstances, ZnO NPs caused an increase in the total amount of proline, glycine betaine, free amino acids, and carbohydrates. In addition, foliar application of ZnO NPs was shown to be effective in preventing the drought-induced decrease in the amount of phenol and mineral nutrients [67]. Stimulatory effects were seen on total chlorophyll (0.93 g/kg FW), shoot biomass (2168 kg/ha), and essential oil content (3.4 g/kg) and yield (7.4 kg/ha) as a result of sub-risky levels of Zn in nanoform absorption (amounted to an average of 152.6 mg Zn/kg after foliar application of 160 mg of zinc oxide nanoparticles) in comparison with ZnS treatment or control. Foliar application of 160 mg/L ZnO NPs is recommended for optimal micronutrient Zn biofortification, biomass production, and essential oil yield in dragonhead [60]. The researchers used many nanocomposites with different concentrations on different plant species to study their effect on their production and increase the yield [4].

Foliar application of nanoparticles enhanced the foliar nutrient status and crop growth and yield. The nano-enabled foliar application could be an ideal strategy for advancing agricultural productivity. Canola plants had an increase in their biomass after receiving a foliar spray of Ca-NPs at a concentration of 100 mg/L, which was deemed the best dosage. Canola plants had an increase in their biomass after receiving a foliar spray of calcium nanoparticles at a concentration of 100 mg/L, which was deemed the best dosage. Furthermore, Ca-NPs also induced a drought-tolerant response in *Brassica napus* plants, which correlated with an uptick in the expression of key antioxidative defense enzymes (APX, POD, SOD, CAT), secondary metabolites and non-enzymatic components (protease, lipoxygenase, proline, total soluble protein contents, endogenous hormonal biosynthesis) [68]. Grain yield components such as the number of pods per plant and 100-grain weight were also affected by seed priming with nano-Zn. Nano-Zn priming increased grain protein percentage by 21%. Therefore, to increase the yield of white beans, priming treatment with nano-Zn as well as foliar application of zinc + iron can be used [69].

The results demonstrated that nano-K_2SO_4 enhanced shoot dry weight, plant height, number of flowers, number of tillers, root length, root fresh weight, and root dry weight under both salt levels on two alfalfa (*Medicago sativa* L.) genotypes [70].

Currently, the use of nanoparticles is having an impact on agricultural production. There is evidence that the results obtained indicated that the foliar application of CuNPs improved the physical and nutraceutical quality and the concentration of Cu in melon fruits. The highest weight and the best diameters of the fruit were obtained with copper nanoparticles have a strong impact on the growth and development of different crops. Biofortification specifically with Cu NPs improves the nutritional quality of food and its consumption has a positive influence on the health of humanity. The findings indicate that there is a statistically significant relationship between the copper nanoparticles and the phytochemical variables found in melon fruits. It was determined that the use of Cu NPs may be an option to enrich melon fruits, and it has the potential to alleviate the problem of copper insufficiency in the diet of the general population [71].

Nano-fertilizers suggest new crop management strategies. Although potassium (K) is difficult to incorporate into organic materials, it helps to increase rice crop quality [72]. Although the dosage of 200 ppm of K nanofertilizer resulted in a greater increase in yield, the dosage of 100 ppm produced greater accumulations of biomass, total chlorophyll content, SPAD values, photosynthetic activity, and nitrate reductase activity. Based on the findings obtained, it seems that the use of K nanofertilizers has a positive impact on the physiological growth of plants [73]. Drip irrigation with 75% nN and foliar treatment with 25% nN have a substantial effect on growth and biochemical parameters [74].

Plant species vary in the efficiency of nanoparticle absorption and the impact of nanoparticles on growth and metabolic processes. Nanoparticle concentration influences activities such as germination and plant growth and development [66].

Various kinds of nanomaterials have shown great promise in promoting sustainable agriculture as they help to improve agricultural production by increasing the efficiency of inputs and minimizing yield losses. Nanomaterials offer a wider specific surface area to fertilizers [75].

Our most significant findings are as follows: (1) Plants have the ability to take up nanomaterials *via* their roots and leaves, which will then subsequently be transformed by the plant. (2) Plant growth may be stimulated and stress alleviated with moderate applications, while high concentrations can be harmful. (3) It is impossible to ignore the impact that nanomaterials have on the rhizosphere [76].

Using nanofertilizers is a potential strategy that may boost the sustainability and efficiency of the agricultural output of farmed crops. This technology is also effective. This technology is also successful due to its nano-size features, high nutrient usage efficiency, gradual release of nutrients, and therefore low necessary applied dosage of fertilizer [77].

Smart fertilizer means using smart agro-technological and advanced tools for the control dose and time of applied fertilizers such as global positioning systems, and remote sensing. These tools are able to minimize agrochemical inputs and maximize crop yield [63].

Nano-fertilization is considered an emerging strategy for increasing plant production while avoiding agroecosystem contamination.

One way to boost agricultural efficiency is the use of nanotechnology. At present, numerous novel nanomaterials are commercially used in agriculture and developed to improve crop productivity and preserve food quality and safety [78].

2.4 The relationship between nanofertilizer and stresses

Many scientists have studied the effect of factors that threaten crop production, including water stress, salinity, and alkalinity, among others. Nanotechnology is a promising tool for increasing crop yields and improving plant stress tolerance. Many books and chapters talk about one or more types of stress. Among them, for example, is a book, Nanomaterial Interactions with Plant Cellular Mechanisms and Macromolecules and Agricultural Implications, and accordingly we tried to start from where the others ended, as we limited ourselves to presenting some research that was published in 2022 and beyond.

Table 1 shows the type of stresses as well as the impact of nanoparticles by different concentrations on the growth and development of various plants.

3. The role of nanofertilizer in environmental protect

The essential key to agricultural sustainability is to perform all farming activities in a sustainable manner [91]. The agricultural sector is facing a series of biological and environmental challenges [92]. Therefore, it is essential to manage the available natural resources in a sustainable method [93]. This requires finding more creative and innovative solutions using modern technology such as nanotechnology that can provide society with the proper technologies used in environmental detection, sensing, and remediation [94]. Farmers face several challenges in traditional agriculture. There are no fertilizers can successfully provide optimal plant nutrients. In addition, chemical toxicity results from the high use of fungicides/pesticides/herbicides [95]. The chemical soluble fertilizers used not only have less utilization efficiency (less than 30%) but large amounts of these are also lost during the application causing environmental pollution [96].

The use of nanoparticles would decrease the consumption of agrochemicals due to the very high efficiency. The environmental benefit of using nanoparticles is more known when it starts to be used, for example, in plant protection products, chemical copper could be replaced with a suitable Cu-NP which is the least environmentally harmful [97]. Likewise, the seed soaking with nanoparticles solution has shown improvements toward increased and more stable production [98]. In order to use nanoparticles in a clear legislative framework, it is necessary to synthesize them with an innovative composition as well as similar in their properties as much as possible to natural substances [99]. In this context, nanofertilizers synthesized by the bio-method are promising substances for achieving sustainable growth of crop yields and securing food for the ever-increasing population [79].

The soil is a system full of life, particularly symbiotic associations with the roots of different plant species [99]. Plants exchange gas and fluid with the environment *via* their leaves and roots, where rhizosphere microorganisms (rhizobacteria and mycorrhizal fungi) solubilize minerals and then the plants absorb solubilize nutrients [100]. This can reflect on the germination of seeds, root and shoot development, biomass, and yield of the crops [101].

DOI: http://dx.doi.org/10.5772/intechopen.1001441

Stresses	Plant	Nanoparticles	Concentration	Applications	Reference
Salinity	Pea (*Pisum sativum* L.)	Si and NSi	3 mM	Enhanced the following indications as vegetative growth, relative water content (RWC), plant height, fresh dry weight, total yield, antioxidant defense systems, and K^+ content in roots and shoots	[79]
	Rice (*Oryza sativa* L.)	ZnO-NPs	50 mg/L	Root length, root fresh weight, root dry weight, root K^+ content, and root antioxidant enzymatic activity were all enhanced by applying 50 mg/L ZnO-NPs often in salinity as well as improved the K^+/Na^+ ratio in the rice's root system.	[80]
	Squash (*Cucurbita pepo* L.)	Fertilizer (nano-K)	0.50 g/L nano-K	Using nano-K to promote antioxidant and photo-synthetic machinery, minimize oxidative stress biomarkers and Na^+ levels, boost tolerance to salt stress, and improve squash yield and yield quality under salt stress.	[81]
	Spearmint (*Mentha spicata* L.)	Chitosan-melatonin nanoparticles (CTS-HPMC-Mel NPs)	an chemical priming agent	Adverse effects of salt stress were ameliorated with Mel and CTS-HPMC-Mel NP treatments by enhancing morphological traits, proline, antioxidant enzymatic activities, as well as the content of dominant constituents of essential oil profile. Engineered CTS-HPMC-Mel NPs could be applied as an innovative protective agent to mitigate the effects of salinity in crop plants.	[82]
Drought	Rice (*Oryza sativa* L.)	ZnONPs	25 ppm	increase both of plant height, total chlorophyll contents, plant fresh and dry weights as well as seed, straw yield and the 1000 paddy weight of rice plants	[83]
	Wheat (*Triticum aestivum* L.)	Iron nanoparticles 10 mg/L when combined with *Glomus in- traradices*	10 mg/L when combined with *Glo-musint-raradices*	Iron nanoparticles combined with *Glomus intraradices* promoted significantly the growth and drought-tolerant of wheat	[84]
	Corn (*Zea mays* L.)	(CNPs)	4.5, 10 nm	Increase in their anti-oxidant defense system.	[40]
	Peppermint (*Mentha piperita* L.)	Myco-Root + TiO_2 NPs	100 mg/L	The maximum content of menthol, 1,8-cineole, and neo-menthol was obtained under mild drought stress (I40) fertilized with Myco-Root + TiO_2 NPs.	[85]

Stresses	Plant	Nanoparticles	Concentration	Applications	Reference
Potentially toxic elements (PTEs)	Barley (*Hordeum vulgare* L.)	nano-Fe_2O_3	1% (w/w)	Lower inhibitory effects on biometric parameters as well as lowering PTEs-induced oxidative damage and protect the growth of barley plants under contaminated soils.	[86]
Cd stress	alfalfa supplemented (*Medicago sativa*)	nZnO	Soil application of 90 mg/kg nZnO with BC (2%)	Decreased Cd and increased Zn-bioaccumulation into roots and shoots as well as higher nZnO and BC levels efficiently alleviated the Cd-mediated reductions in alfalfa biomass, anti-oxidant enzymatic response, and gaseous exchange traits than control.	[87]
Soils contaminated with rare earth elements (REEs)	Duckweed (*Lemna minor*)	nano terbium (Tb)	less than 100 mg/L	Increased the contents of (N), (P), (K^+), (Ca^{+2}), (Mg^{+2}), (Mn^{+2}) and (Fe^{+2})	[88]
CeO_2	Pakchoi (*Brassica chinensis* L.)	CeO_2 NPs	0.7 mg/kg	Sub-stomatal CO_2 was increased dramatically under low doses of CeO_2 NPs	[89]
Cu stress	Rape (*Brassica napus* L.)	SNPs	5 mg/L	SNPs application enhanced the shoot height, root length, and dry weight of shoot and root by 34.6%, 282%, 41.7% and 37.1%, respectively, over Cu treatment alone SNPs application has the sustainable technology for increasing plant productivity and reducing the accumulation of toxic metals in heavy metal-polluted soils.	[90]

Table 1.
The various impacts of different types of stresses with different concentrations on growth and development of plants.

Nanofertilizers have some characteristics compared to conventional fertilizers [99]. These features help crops absorb nutrients slowly and sustainably due to a high surface-to-volume ratio and reduced loss of nutrients [102]. Also, it is having active sites for a more significant number of biological activities, which increases the efficiency of plant absorption of nanofertilizer [103]. Furthermore, soil fertility improvement as well as maintaining a suitable environment for the growth of microorganisms in the soil [104]. Thus, it is providing sustainable solutions to environmental pollution and climate change [105].

Methods for preparing nanoparticles vary to the available capabilities and the purpose of the application [106]. Many of the protocols relied on eco-friendly, less polluting, more sustainable approaches [107]. Also, it can use extraction and separation strategies away from pressures [108], high temperatures [109], acidifications [110], and toxic metals [111]. The most natural sources used to synthesize nanoparticles from plant extracts as well as microorganisms such as yeasts, fungi, bacteria, and algae. Biomolecules of cells microorganisms and plants such as enzymes, amines, proteins, phenolic, alkaloids, and pigments contribute to synthesizing NPs and capping of them [112].

Bioremediation provides based on nanomaterials many features compared to conventional treatments. It has exhibited a quantum effect with decreased activation energy as well as contact with a larger amount of it with the surrounding materials, and also, has a high competence level, selective to specific metals, and is more economically feasible [113].

Bioremediation processes were used in groundwater and wastewater management, uranium remediation, solid waste, soil remediation and remediation of heavy metal pollution as well as petroleum [114]. The ability of NMs to combat pollution is a revolutionary change in the ecological field. Furthermore, it can mention these changes *via* some examples such as the nanoscale zero-valent iron removing AS (III) in anoxic groundwater [115]. Eliminating hydrophobic contaminants was *via* using engineered polymeric NPs [116]. Formatting special structures of PAMAM dendrimers are used in water treatment [117]. Employment-engineered polymeric NPs are used for soil remediation [118].

On other hand, NPS have a great effect on biodegradation especially dye contaminants. These dyes have high chemical stability, and compositions complicated and long-distance stability in running water, which will lead to inhibited photosynthesis, the development of aquatic biota and less use of dissolved oxygen as well as decreased watercourse recovery rate **Figure 1** [119].

In addition, several studies showed the importance of zinc oxide (ZnO) and titanium dioxide (TiO_2) in environmental remediation which were used in dye-sensitized solar cells as well as water photoelectrolysis [120].

NPs have the best-advanced strategies for the treatment of wastewater due to their properties such as high adsorption and reaction capacities as well as a suspension in aqueous solutions as colloids, less time-consuming, efficient cost-effective, eco-friendly and less waste compared with conventional methods [121].

Green nanomaterials have a wide effective of abilities to remove organic and inorganic solutes, toxic metal ions and pathogenic microorganisms from water by using various nanomaterials such as nanosorbents (carbon-based nanosorbents such as Captymer™), nanostructured catalytic membranes (cellulose acetate, polyvinylidene fluoride, chitosan and polysulfone), nanocatalysts (silver nanocatalysts), bioactiveNPs, molecularly imprinted polymers (MIPs) and biomimetic membranes [122, 123].

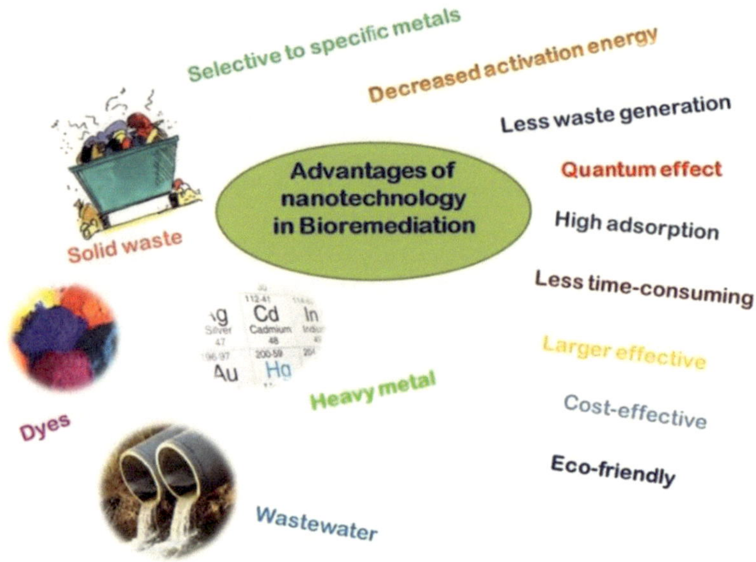

Figure 1.
Graphic representation of the advantages of nanotechnology in bioremediation.

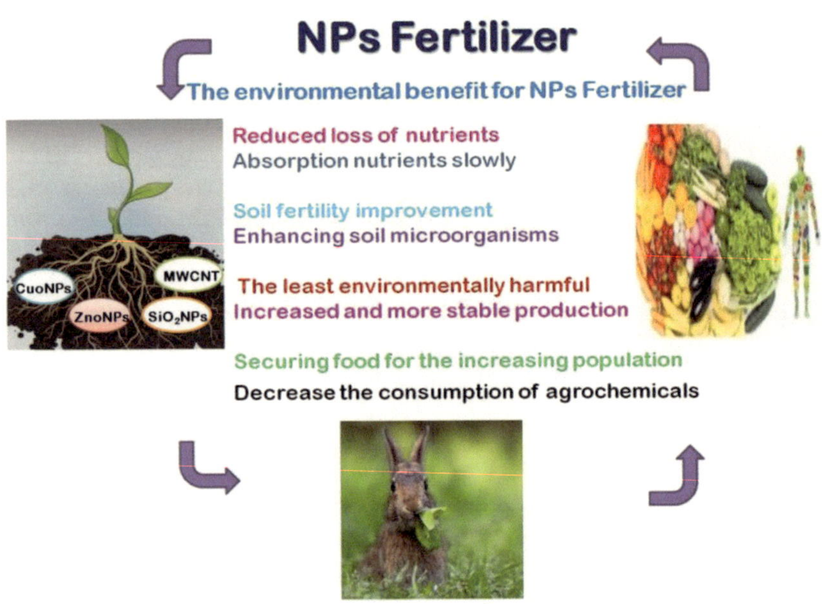

Figure 2.
Graphic representation of the environmental benefit for nanoparticles fertilizer.

4. Nanofertilizer toxicity

NPs found naturally in soil *via* agricultural activities or rain as well as through anthropogenic production [124]. Also, it is more concentrated in soil compared to air and water. Plants absorb these NPs from the soil and move them to reach various parts of plants and finally, a human consumes this plant **Figure 2** [125].

The NPs' increasing application in agriculture led to directly interacting with the environment as well as harmful effects on aquatic, terrestrial species and humans [105]. It is reflected in the accumulation of NPs in the soil, fertility and physicochemical characteristics of the soil [126].

The NPs' harmful effects on plants include DNA damage, reactive oxygen species (ROS) formation, association with nuclear protein, chromosomal aberration, decreased DNA repair and genetic defects. NiO NPs access the DNA of tomato plants and produced irreversible cell damage [127, 128]. Application Co_3O_4 NPs in eggplants caused apoptosis in their cells [128]. In addition, using ZnO NPs led to adverse effects on membrane integrity, chromosomal damage and DNA strand breakage in *Vicia faba* L., *Nicotiana tabacum* L. and *Allium cepa* L. [129, 130].

The presence of silver NPs in the range of 0.1–0.5 mg kg − 1 in soil inhibits the growth of dehydrogenase activity of bacteria [131].

Likewise, CNTs, copper NPs, zinc NPs and iron NPs have several adverse effects on soil microflora and aquatic life, which include aquatic microbes, vertebrates and aquatic plants [132, 133]. It is reported [134] that graphene oxide NPs caused toxic effects on freshwater algae due to the generation of oxidative stress.

5. Conclusion and future perspective

Sustainable development to achieve the Zero Hunger SDG 2030 should be inclusive. This goal is achieved *via* the implementation of agricultural practices to ensure improved crop production, protecting the natural resources (water, land and forests) as well without adverse impacts on biota and human beings [105].

Nanotechnology can achieve these requirements by relying on vital nanomaterials in environmental remediation and other industries. It is cost effective and involves less toxic waste disposal. Also, it improves catalytic efficiency and selectivity.

Despite the fact that the reports that are readily available on the application of nanoparticles to the process of cultivating soil and plants are only for a small number of nanomaterial, the quantity of materials used in agriculture is significantly greater than what is currently available.

As a result of this, there is an urgent need to implement a bigger variety of nanomaterials, each of which must be manufactured using a new precursor and creative synthesized methods. In addition, using plant components in the manufacturing of nanoparticles during green synthesis may result in the development of nanoparticles that are well suited for performing effectively as nanofilters.

However, the growing application of NPs must be a concern on some points, their accumulation in the soil, their interaction with plants and microorganisms and the dose of application as well as toxicological aspects in ecotoxicological and food chains.

The potential impact of nanofertilizers on human health necessitates more investigation into their persistence in plant tissue that is consumed. While nanofertilizers have the potential to radically alter farming methods, it is essential to evaluate their effects on the environment as soon as possible.

Therefore, our insight will concern the development of green strategies for biogenic nanoparticle synthesis, examining plant/soil–NP interactions in the field in a way that enhances crop productivity and provides a suitable environment for the growth and development of all living organisms and feeding future generations with healthy food.

Author details

Lina M. Alnaddaf[1*], Salim F. Bamsaoud[2] and Mahroos Bahwirth[3]

1 Biotechnology and Molecular Biology, Department of Field Crops, College of Agriculture, Albaath University, Homs, Syria

2 Faculty of Science, Physics Department, Center for Natural and Applied Science, Hadhramout Foundation for Invention and Advancement of Sciences, Hadhrmaout University, Mukalla, Yemen

3 Faculty of Science, Biology Department, Center for Natural and Applied Science, Hadhramout Foundation for Invention and Advancement of Sciences, Hadhramout University, Mukalla, Yemen

*Address all correspondence to: hatem005@gmail.com

References

[1] Omar RA, Afreen S, Talreja N, Chauhan D, Ashfaq MOMAR, Anand R, et al. Impact of nanomaterials in plant systems. In: Plant Nanobionics. Cham: Springer; 2019. pp. 117-140

[2] Rajput VD, Minkina TM, Behal A, Sushkova SN, Mandzhieva S, Singh R, et al. Effects of zinc-oxide nanoparticles on soil, plants, animals and soil organisms: A review. Environmental Nanotechnology, Monitoring & Management. 2018;**9**:76-84. DOI: 10.1016/j.enmm.2017.12.006

[3] Liu R, Lal R. Potentials of engineered nanoparticles as fertilizers for increasing agronomic productions. Science of the Total Environment. 2015;**1**(514):131-139. DOI: 10.1016/j.scitotenv.2015.01.104 Epub 2015 Feb 5

[4] Bahwirth MA, Bamsaoud SF, Alnaddaf LM. Nanomaterial impact on plant morphology, physiology and productivity. In: Al-Khayri JM, Alnaddaf LM, Jain SM, editors. Nanomaterial Interactions with Plant Cellular Mechanisms and Macromolecules and Agricultural Implications. Cham: Springer; 2023. pp. 319-340. DOI: 10.1007/978-3-031-20878-2_12

[5] Ponder SM, Darab JG, Bucher J, Caulder D, Craig I, Davis L, et al. Surface chemistry and electrochemistry of supported zerovalent iron nanoparticles in the remediation of aqueous metal contaminants. Chemistry of Materials. 2001;**13**(2):479-486. DOI: 10.1021/cm000288r

[6] Lowry GV, Johnson KM. Congener-specific dechlorination of dissolved PCBs by microscale and nanoscale zerovalent iron in a water/methanol solution. Environmental Science & Technology. 2004;**38**(19):5208-5216. DOI: 10.1021/es049835q

[7] Liou YH, Lo SL, Lin CJ, Kuan WH, Weng SC, Liou YH, et al. Chemical reduction of an unbuffered nitrate solution using catalyzed and uncatalyzed nanoscale iron particles. Journal of Hazardous Materials. 2005;**127**(1-3):102-110. DOI: 10.1016/j.jhazmat.2005.06.029 Epub 2005 Aug 2. PMID: 16081210

[8] Sohn K, Kang SW, Ahn S, Woo M, Yang SK. Fe(0) nanoparticles for nitrate reduction: Stability, reactivity, and transformation. Environmental Science & Technology. 2006;**40**(17):5514-5519. DOI: 10.1021/es0525758

[9] Liu Y, Phenrat T, Lowry GV. Effect of TCE concentration and dissolved groundwater solutes on NZVI-promoted TCE dechlorination and H2 evolution. Environmental Science & Technology. 2007;**41**(22):7881-7887. DOI: 10.1021/es0711967 PMID: 18075103

[10] Gupta N, Ram H, Kumar B. Mechanism of zinc absorption in plants: Uptake, transport, translocation and accumulation. Reviews in Environmental Science and Bio/Technology. 2016;**15**(1):89-109. DOI: 10.1007/s11157-016-9390-1

[11] Abbas HH, Noufal E, Farid I, Attia M, Ahmed R, Abbas M. Effect of traditional sources of Zn and ZnO-nano-particles foliar application on productivity and P-uptake of maize plants grown on sandy and clay loam soils. Environment, Biodiversity and Soil Security. 2021;**5**:59-72. DOI: 10.21608/jenvbs.2021.58802.1127

[12] Ali NM, Altaey DK, Altaee NH. The impact of selenium, Nano (SiO_2)

and organic fertilization on growth and yield of potato *Solanum tuberosum* L. under salt stress conditions. In: IOP Conf. Ser.: Earth Environ. Sci. Vol. 735. IOP Publishing; 2021. p. 012042. DOI: 10.1088/1755-1315/735/1/012042

[13] He Y, Hu R, Zhong Y, Zhao X, Chen Q, Zhu H. Graphene oxide as a water transporter promoting germination of plants in soil. Nano Research. 2018;**11**(4):1928-1937. DOI: 10.1007/s12274-017-1810-1

[14] Shah V, Collins D, Walker VK, Shah S. The impact of engineered cobalt, iron, nickel and silver nanoparticles on soil bacterial diversity under field conditions. Environmental Research Letters. 2014;**9**(2):024001. DOI: 10.1088/1748-9326/9/2/024001

[15] Wei X, Cao P, Wang G, Liu Y, Song J, CuO HJ. ZnO, and γ-Fe2O3 nanoparticles modified the underground biomass and rhizosphere microbial community of salvia miltiorrhiza (Bge.) after 165-day exposure. Ecotoxicology and Environmental Safety. 2021;**1**(217):112232. DOI: 10.1016/j.ecoenv.2021.112232 Epub 2021 Apr 14

[16] Suriyaprabha R, Karunakaran G, Yuvakkumar R, Rajendran V, Kannan N. Silica nanoparticles for increased silica availability in maize (Zea mays. L) Seeds under hydroponic conditions. Current Nanoscience. 2012;**8**(6):902-908. DOI: 10.2174/157341312803989033

[17] Karunakaran G, Suriyaprabha R, Manivasakan P, Yuvakkumar R, Rajendran V, Prabu P, et al. Effect of nanosilica and silicon sources on plant growth promoting rhizobacteria, soil nutrients and maize seed germination. IET Nanobiotechnology. 2013;**7**(3):70-77. DOI: 10.1049/iet-nbt.2012.0048

[18] Liu J, Cai H, Mei C, Wang M. Effects of nano-silicon and common silicon on lead uptake and translocation in two rice cultivars. Frontiers of Environmental Science & Engineering. 2015;**9**(5):905-911. DOI: 10.1007/s11783-015-0786-x

[19] Tantawy AS, Salama YAM, El-Nemr MA, Abdel-Mawgoud AMR. Nano silicon application improves salinity tolerance of sweet pepper plants. International Journal of ChemTech Research. 2015;**8**(10):11-17

[20] Tripathi DK, Singh S, Singh VP, Prasad SM, Chauhan DK, Dubey NK. Silicon nanoparticles more efficiently alleviate arsenate toxicity than silicon in maize cultiver and hybrid differing in arsenate tolerance. Frontiers in Environmental Science. 2016;**4**:46. DOI: 10.3389/fenvs.2016.00046

[21] Soares C, Branco-Neves S, de Sousa A, Teixeira J, Pereira R, Fidalgo F. Can nano-SiO2 reduce the phytotoxicity of acetaminophen? - A physiological, biochemical and molecular approach. Environmental Pollution. 2018;**241**:900-911. DOI: 10.1016/j.envpol.2018.06.037

[22] Ayman M, Metwally S, Mancy M, Abd Alhafez A. Influence of nano–silica on wheat plants grown in salt–affected soil. Journal of Productivity and Development. 2020;**25**(3):279-296. DOI: 10.21608/JPD.2020.120786

[23] Simonin M, Martins JM, Le Roux X, Uzu G, Calas A, Richaume A. Toxicity of TiO2 nanoparticles on soil nitrification at environmentally relevant concentrations: Lack of classical dose-response relationships. Nanotoxicology. 2017;**11**(2):247-255. DOI: 10.1080/17435390.2017.1290845

[24] Juárez-Maldonado A, Ortega-Ortíz H, Morales-Díaz AB, et al.

Nanoparticles and nanomaterials as plant biostimulants. International Journal of Molecular Sciences. 2019;**20**(1):162. DOI: 10.3390/ijms20010162

[25] Thiagarajan V, Ramasubbu S. Fate and behaviour of TiO_2 nanoparticles in the soil: Their impact on staple food crops. Water, Air, & Soil Pollution. 2021;**232**(7):1-18. DOI: 10.1007/s11270-021-05219-8

[26] Mohammadi R, Maali-Amiri R, Mantri NL. Effect of TiO_2 nanoparticles on oxidative damage and antioxidant defense systems in chickpea seedlings during cold stress. Russian Journal of Plant Physiology. 2014;**61**(6):768-775. DOI: 10.1134/S1021443714050124

[27] Qi M, Liu Y, Li T. Nano-TiO_2 improve the photosynthesis of tomato leaves under mild heat stress. Biological Trace Element Research. 2013;**156**(1-3):323-328. DOI: 10.1007/s12011-013-9833-2

[28] Jaberzadeh A, Moaveni P, Moghadam HRT, Zahedi H. Influence of bulk and nanoparticles titanium foliar application on some agronomic traits, seed gluten and starch contents of wheat subjected to water deficit stress. Notulae botanicae horti agrobotanici cluj-napoca. 2013;**41**(1):201-207. DOI: 10.15835/nbha4119093

[29] Mohammadi H, Esmailpour M, Gheranpaye A. Effects of TiO_2 nanoparticles and water-deficit stress on morpho-physiological characteristics of dragonhead (*Dracocephalum moldavica* L.) plants. Acta Agriculturae Slovenica. 2016;**107**(2):385-396. DOI: 10.14720/aas.2016.107.2.11

[30] Kiapour H, Moaveni P, Habibi D. Evaluation of the application of gibbrellic acid and titanium dioxide nanoparticles under drought stress on some traits of basil (*Ocimum basilicum* L.). IJAAR. 2015, 2015;**6**(4):138-150

[31] Khan MN. Nano-titanium dioxide (nano-TiO_2) mitigates NaCl stress by enhancing antioxidative enzymes and accumulation of compatible solutes in tomato (Lycopersicon esculentum mill.). Journal of Plant Sciences. 2016;**11**(1/3):1-11. DOI: 10.3923/jps.2016.1.11

[32] Kaur H, Kalia A, Sandhu JS, Dheri GS, KaurG, & Pathania S. Interaction of TiO_2 nanoparticles with soil: Effect on microbiological and chemical traits. Chemosphere. 2022;**301**:134629

[33] Khana MA, Khanb T, Mashwanic ZR, et al. Plant cell nanomaterials interaction: Growth, physiology and secondary metabolism. In: Comprehensive Analytical Chemistry. Vol. 84. Amsterdam: Elsevier; 2019. pp. 32-54. DOI: 10.1016/bs.coac.2019.04.005

[34] Cañas JE, Long M, Nations S, et al. Effects of functionalized and nonfunctionalized single-walled carbon nanotubes on root elongation of select crop species. Environmental Toxicology and Chemistry. 2008;**27**(9):1922-1931. DOI: 10.1897/08-117.1

[35] Lahiani MH, Dervishi E, Chen J, et al. Impact of carbon nanotube exposure to seeds of valuable crops. ACS Applied Materials & Interfaces. 2013;**5**(16):7965-7973. DOI: 10.1021/am402052x

[36] Jiang Y, Hua Z, Zhao Y, Liu Q, Wang F, Zhang Q. The effect of carbon nanotubes on rice seed germination and root growth. In: Proceedings of the 2012 International Conference on Applied Biotechnology (ICAB 2012). Berlin, Heidelberg: Springer; 2014. pp. 1207-1212. DOI: 10.1007/978-3-642-37922-2_129

[37] Martínez-Ballesta MC, Zapata L, Chalbi N, Carvajal M. Multiwalled carbon nanotubes enter broccoli cells enhancing growth and water uptake of plants exposed to salinity. Journal of Nanobiotechnology. 2016;**14**(1):42. DOI: 10.1186/s12951-016-0199-4

[38] Zhao F, Xin X, Cao Y, et al. Use of carbon nanoparticles to improve soil fertility, crop growth and nutrient uptake by corn (*Zea mays* L.). Nanomaterials (Basel). 2021;**11**(10):2717. DOI: 10.3390/nano11102717

[39] Xin X, Nepal J, Wright AL, Yang X, He Z. Carbon nanoparticles improve corn (*Zea mays* L.) growth and soil quality: Comparison of foliar spray and soil drench application. Journal of Cleaner Production. 2022;**363**:132630. DOI: 10.1016/j.jclepro.2022.132630

[40] Alsherif EA, Almaghrabi O, Elazzazy AM, Abdel-Mawgoud M, Beemster GTS, Abd EH. Carbon nanoparticles improve the effect of compost and arbuscular mycorrhizal fungi in drought-stressed corn cultivation. Plant Physiology and Biochemistry. 2023;**194**:29-40. DOI: 10.1016/j.plaphy.2022.11.005 Epub 2022 Nov 10

[41] Singh A, Singh NB, Hussain I, Singh H, Singh SC. Plant-nanoparticle interaction: An approach to improve agricultural practices and plant productivity. International Journal of Pharmaceutical Science Invention. 2015;**4**(8):25-40

[42] Aslani F, Bagheri S, Muhd Julkapli N, Juraimi AS, Hashemi FS, Baghdadi A. Effects of engineered nanomaterials on plants growth: An overview. The Scientific World Journal. 2014;**2014**:641759. DOI: 10.1155/2014/641759

[43] Remedios C, Rosario F, Bastos V. Environmental nanoparticles interactions with plants: Morphological, physiological, and genotoxic aspects. Journal of Botany. 2012;**2012**:1-8. DOI: 10.1155/2012/751686

[44] Predoi D, Ghita RV, Iconaru SL, et al. Application of Nanotechnology Solutions in Plants Fertilization, Urban Horticulture - Necessity of the Future. London, UK, London: IntechOpen; 2020. pp. 51-74. DOI: 10.5772/intechopen.91240

[45] Hossain Z, Yasmeen F, Komatsu S. Nanoparticles: Synthesis, morphophysiological effects, and proteomic responses of crop plants. International Journal of Molecular Sciences. 2020;**21**:3056-3064. DOI: 10.3390/ijms21093056

[46] Velasco EAP, Galindo RB, Aguilar LAV, et al. Effects of the morphology, surface modification and application methods of ZNO-NPS on the growth and biomass of tomato plants. Molecules. 2020;**25**:1282. DOI: 10.3390/molecules25061282

[47] Itroutwar PD, Govindaraju K, Tamilselvan S, et al. Seaweed-based biogenic zno nanoparticles for improving agro-morphological characteristics of rice (Oryza sativa L.). Journal of Plant Growth Regulation. 2019;**39**:717-728. DOI: doi.org/10.1007/s00344-019-10012-3

[48] Bamsaoud S, Bahwirth M. The effect of biologically synthesized silver nanoparticles on the germination and growth of *Cucurbita pepo* seedlings. Journal of the Arab American University. 2017;**3**:34-47

[49] Bahwirth M, Bamsaoud S. The effect of silver nanoparticles on the germination and growth of two cultivars

of wheat *Triticum aestivum* L. Alandalus Journal for Applied Sciences. 2020;7:9-25

[50] Al-Hadede LT, Khaleel SA, Hasan SK. Some applications of nanotechnology in agriculture. Biochemical and Cellular Archives. 2020;**20**:1447-1454. DOI: 10.35124/bca.2020.20.1.1447

[51] Bamsaoud S, Basuliman M, Bin-Hameed A, et al. The effect of volume and concentration of AgNO3 aqueous solutions on silver nanoparticles synthesized using Ziziphus Spina–Christi leaf extract and their antibacterial activity. Journal of Physics: Conference Series. 2021;**1900**(1):012005

[52] Siddiqui H, Ahmed KBM, Sami F, Hayat S. Silicon nanoparticles and plants: Current knowledge and future perspectives. In: Hayat S et al., editors. Sustainable Agriculture Reviews. Vol. 41. Cham: Springer; 2020

[53] Elshayb OM, Nada AM, Ibrahim HM, et al. Application of silica nanoparticles for improving growth, yield, and enzymatic antioxidant for the hybrid rice EHR1 growing under water regime conditions. Materials. 2021;**14**:1150-1169. DOI: 10.3390/ma14051150

[54] Elemike EE, Uzoh IM, Onwudiwe DC, Babalola OO. The role of nanotechnology in the fortification of plant nutrients and improvement of crop production. Applied Sciences. 2019;**9**:499-532. DOI: 10.3390/app9030499

[55] Guo Y, Khan RAA, Xiong Y, Fan Z. Enhanced suppression of soil-borne phytopathogenic bacteria Ralstonia solanacearum in soil and promotion of tomato plant growth by synergetic effect of green synthesized nanoparticles and plant extract. Journal of Applied Microbiology. 2022;**132**(5):3694-3704

[56] Azmat A, Tanveer Y, Yasmin H, Hassan MN, Shahzad A, Reddy M, et al. Coactive role of zinc oxide nanoparticles and plant growth promoting rhizobacteria for mitigation of synchronized effects of heat and drought stress in wheat plants. Chemosphere. 2022;**297**:133982

[57] Dappula SS, Kandrakonda YR, Shaik JB, Mothukuru SL, Lebaka VR, Mannarapu M, et al. Biosynthesis of zinc oxide nanoparticles using aqueous extract of Andrographis alata: Characterization, optimization and assessment of their antibacterial, antioxidant, antidiabetic and anti-Alzheimer's properties. Journal of Molecular Structure. 2023;**1273**:134264

[58] Wasilewska A, Klekotka U, Zambrzycka M, Zambrowski G, Święcicka I, Kalska-Szostko B. Physico-chemical properties and antimicrobial activity of silver nanoparticles fabricated by green synthesis. Food Chemistry. 2023;**400**:133960

[59] Tiwari K, Kumar Y, Singh T, Nayak R. Nano technology based P fertilizers for higher efficiency and agriculture sustainability. Annals of Plant and Soil Research. 2022;**24**(2):198-207. DOI: 10.47815/apsr.2022.10149

[60] Reddy SS, Chhabra V. Nanotechnology: Its scope in agriculture. Journal of Physics: Conference Series. 2022;**2267**:012112. DOI: 10.1088/1742-6596/2267/1/012112

[61] Marjan N, Sina F, Ali A, Surki L, Raj P, Ali R. Improved efficacy of foliar application of zinc oxide nanoparticles on zinc biofortification, primary productivity and secondary metabolite production in dragonhead. Journal of Cleaner Production. 2022;**379**:379. DOI: 10.1016/j.jclepro.2022.134803

[62] Verma K, Song P, Joshi A, Tian D, Rajput D, Singh M, et al. Recent trends in Nano-Fertilizers for sustainable agriculture under climate change for global food security. Nanomaterials (Basel, Switzerland). 2022;**12**(1):173. DOI: 10.3390/nano12010173

[63] Aguirre H, Feregrino A, Esquivel P, Vazquez C, Mariana A. Nanomaterials as an alternative to increase plant resistance to abiotic stresses. Frontiers in Plant Science. 2022;**13**:1023636. DOI: 10.3389/fpls.2022.1023636

[64] Abdalla Z, El-Sawy S, El-Bassiony M, Jun H, Shedeed S, Okasha A, et al. Nano-fertilizers: A pictorial overview. Environment, biodiversity and soil. Security. 2022;**6**:191-204. DOI: 10.21608/jenvbs.2022.153990.1184

[65] Guha T, Mukherjee A, Kundu R. Nano-scale zero Valent iron (nZVI) priming enhances yield, alters mineral distribution and grain nutrient content of Oryza sativa L. cv. Gobindobhog: A Field study. Journal of Plant Growth Regulation. 2022;**41**:710-733. DOI: 10.1007/s00344-021-10335-0

[66] Beig B, Niazi K, Sher F, Bilal B, Muhammad N, Farooq S, et al. Nanotechnology-based controlled release of sustainable fertilizers A review. Environmental Chemistry Letters. 2022;**20**:2709-2726. DOI: 10.1007/s10311-022-01409-w

[67] Hayyawi W, Al-Juthery A, Sana Y, Nabil L, Ali A, Sulaf T, et al. Response wheat to spray some of synthetic Nano Fertilizers. IOP Conference Series: Earth and Environmental Science. 2022;**1060**:012030. DOI: 10.1088/1755-1315/1060/1/012030

[68] Ghani MI, Saleem S, Rather SA, Rehmani MS, Alamri S, Rajput VD, et al. Foliar application of zinc oxide nanoparticles: An effective strategy to mitigate drought stress in cucumber seedling by modulating antioxidant defense system and osmolytes accumulation. Chemosphere. 2022;**289**:133202. DOI: 10.1016/j.chemosphere.2021.133202

[69] Ahsan A, Rouyi F, Junyi M, Fakhir H, Qian H, Habib A, et al. Calcium nanoparticles (Ca-NPs) improve drought stress tolerance in Brassica napus by modulating the photosystem II, nutrient acquisition and antioxidant performance. NanoImpact. 2022;**28**:100423. DOI: 10.1016/j.impact.2022.100423

[70] Seyed A, Hamid E, Ahmad I, Naser A. Effect of gibberellin, nano-nutrition with titanium, zinc and iron on yield and some physiological and qualitative traits of white beans. Notulae botanicae Horti Agrobotanici Cluj-Napoca. 2022;**50**(1):1-16. DOI: 10.15835/nbha50112538

[71] El-Shal M, El-Naggar H, El-Beshbeshy R, Mahmoud K, El-Kader A, Missaui M, et al. Effect of Nano-Fertilizers on alfalfa plants grown under different salt stresses in hydroponic system. Agriculture. 2022;**12**(8):1113. DOI: 10.3390/agriculture12081113

[72] Fortis M, Ortiz J, Preciado P, Trejo R, Lagunes E, Andrade A, And Rueda O: Biofortification with copper nanoparticles (Nps Cu) and its effect on the physical and nutraceutical quality of hydroponic melon fruits. Notulae Botanicae Horti Agrobotanici Cluj-Napoca. 2022;**50**, 1:12568. DOI: 10.15835/nbha50112568

[73] Pooja A, Vidyashree S, Shivashankar K, Kiran E, Ashwini R. Effects of nano potassium in rice: A review. Indian Journal of Natural Sciences. 2022;**13**(72):43202-43208. DOI: 10.35709/ory.2021.58.3.1

[74] Ana K, Márquez-Prieto AP, Esteban S, Bertha C, Macias-López S, Octavio VC, et al. Impact of the foliar application of potassium nanofertilizer on biomass, yield, nitrogen assimilation and photosynthetic activity in green beans. Notulae botanicae Horti Agrobotanici Cluj-Napoca. 2022;**50**:1. DOI: 10.15835/nbha50112569

[75] Sharaf-Eldin A, Elsawy B, Eisa Y, El-Ramady H, Usman M, Ziaur-Rehman M. Application of nano-nitrogen fertilizers to enhance nitrogen efficiency for lettuce growth under different irrigation regimes. Pakistan Journal of Agricultural Sciences. 2022;**59**(3):367-379. DOI: 10.21162/PAKJAS/22.1044. Available from: http://www.pakjas.com.pk. ISSN (Online) 2076-09061

[76] Abdul, Qayyum A, Faisal N, Yamin B, Raza U, Husnain R, Husnain JA. Role of Nanoparticles in Improving Stress Tolerance in Crop Plants. CABI Books. CABI International; 2022. DOI: 10.1079/9781789248098.0024

[77] Lingrui L, Niana H, Tengxiang L. Plants and rhizospheric environment: Affected by zinc oxide nanoparticles (ZnO NPs). A review. Plant Physiology and Biochemistry. 2022;**185**:91-100. DOI: 10.1016/j.plaphy.2022.05.032

[78] Nongbet A, Mishra AK, Mohanta YK, Mahanta S, Ray MK, Khan M, et al. Nanofertilizers: A Smart and sustainable attribute to modern agriculture. Plants (Basel). 2022;**11**(19):2587. DOI: 10.3390/plants11192587

[79] Nazari M, Mojtaba K, Akbar G, Jameel A, et al. Role of nanomaterials in improving the nutritional value of crops. In: Nanomaterial Interactions with Plant Cellular Mechanisms and Macromolecules and Agricultural Implications. Cham: Springer; 2023. pp. 399-422

[80] Ismail LM, Soliman MI, Abd El-Aziz MH, Abdel-Aziz HMM. Impact of silica ions and Nano silica on growth and productivity of pea plants under salinity stress. Plants. 2022;**11**(4):494. DOI: 10.3390/plants11040494

[81] Singh A, Sengar RS, Shahi UP, Rajput VD, Minkina T, Ghazaryan KA. Prominent effects of zinc oxide nanoparticles on roots of Rice (*Oryza sativa* L.) grown under salinity stress. Stress. 2023;**3**(1):33-46. DOI: 10.3390/stresses3010004

[82] Rady MM, Mossa A-TH, Youssof AMA, Ashraf S, Osman SMA, Ahmed IAA. Mohamed: Exploring the reinforcing effect of nano-potassium on the antioxidant defense system reflecting the increased yield and quality of salt-stressed squash plants. Scientia Horticulturae. 2023;**308**:111609. DOI: 10.1016/j.scienta.2022.111609

[83] Gholamreza G, Habib F, Sima P, Elnaz Z, Parisa L, Hessam J, et al. Mitigation of salinity impact in spearmint plants through the application of engineered chitosan-melatonin nanoparticles. International Journal of Biological Macromolecules. 2023;**224**:893-907. DOI: 10.1016/j.ijbiomac.2022.10.175

[84] Waqas MM, Ishtiaq M, Hussain I, Parveen A, Hayat Bhatti K, Azeem M, et al. Seed nano-priming with zinc oxide nanoparticles in rice mitigates drought and enhances agronomic profile. PLoS One. 2022;**17**(3):e0264967. DOI: 10.1371/journal.pone.0264967

[85] Naseer M, Zhu Y, Li FM, Yang YM, Wang S, Xiong YC. Nano-enabled improvements of growth and colonization rate in wheat inoculated with arbuscular mycorrhizal fungi. Environmental Pollution. 2022

Feb 15;**295**:118724. DOI: 10.1016/j.envpol.2021.118724 Epub 2021 Dec 21

[86] Ostadi A, Javanmard A, Amani Machiani M, Kakaei K. Optimizing antioxidant activity and phytochemical properties of peppermint (Mentha piperita L.) by integrative application of biofertilizer and stress-modulating nanoparticles under drought stress conditions. Plants. 2023;**12**(1):151. DOI: 10.3390/plants12010151

[87] Rodríguez-Seijo A, Soares C, Ribeiro S, Amil BF, Patinha C, Cachada A, et al. Nano-Fe2O3 as a tool to restore plant growth in contaminated soils - assessment of potentially toxic elements (bio)availability and redox homeostasis in Hordeum vulgare L. Journal of Hazardous Materials. 2022 Mar;**5**(425):127999. DOI: 10.1016/j.jhazmat.2021.127999 Epub 2021 Dec 6

[88] Abdul Kareem H, Adeel M, Azeem M, Ahmad MA, Shakoor N, Hassan MU, et al. Antagonistic impact on cadmium stress in alfalfa supplemented with nano-zinc oxide and biochar via upregulating metal detoxification. Journal of Hazardous Materials. 2023;**443**(Pt B):130309. DOI: 10.1016/j.jhazmat.2022.130309 Epub 2022 Nov 2

[89] Alp FN, Arikan B, Ozfidan-Konakci C, Gulenturk C, Yildiztugay E, Turan M, et al. Hormetic activation of nano-sized rare earth element terbium on growth, PSII photochemistry, antioxidant status and phytohormone regulation in Lemna minor. Plant Physiology and Biochemistry. 2023;**194**:361-373. DOI: 10.1016/j.plaphy.2022.11.031 Epub 2022 Nov 28

[90] Hong J, Jia S, Wang C, Li Y, He F, Gardea-Torresdey JL. Transcriptome reveals the exposure effects of CeO2 nanoparticles on pakchoi (Brassica chinensis L.) photosynthesis. Journal of Hazardous Materials. 2023;**444**(Pt A):130427. DOI: 10.1016/j.jhazmat.2022.130427 Epub 2022 Nov 17

[91] Yuan H, Liu Q, Fu J, Wang Y, Zhang Y, Sun Y, et al. Co-exposure of sulfur nanoparticles and Cu alleviate Cu stress and toxicity to oilseed rape Brassica napus L. Journal of Environmental Sciences (China). 2023;**124**:319-329. DOI: 10.1016/j.jes.2021.09.040 Epub 2022 Feb 3

[92] Ilmudeen S, Farween F, MahaletchumiS, KanchanaDKDI. Advances of Nanofertilizers in modern agriculture. A review. The Journal of Research Technology and Engineering. 2022;**3**(2):19-29

[93] Khan MR, Rizvi TF. Application of nanofertilizer and nanopesticides for improvements in crop production and protection in nanoscience and plant–soil systems. Springer Cham. 2017;**48**:405-427

[94] Ghamry AE, Mosa AA, Alshaal T, Ramady HE. Nanofertilizers vs. biofertilizers: New insights. Environment, Biodiversity and Soil Security. 2018;**2**(2018):51-72

[95] Sahu SK, Sk B, Acharya DK, Padhee RK. Development of nano bio fertilizer via chemical & biological synthesis. IRJMETS. 2022;**4**(7):1-17

[96] Avila-Quezada GD, Ingle AP, Golińska P, Rai M. Strategic applications of nano-fertilizers for sustainable agriculture: Benefits and bottlenecks. De Gruyter Nanotechnology Reviews. 2022;**11**:2123-2140. DOI: 10.1515/ntrev-2022-0126

[97] Zhang Q, Ying Y, Ping J. Recent advances in plant nanoscience. Advancement of Science. 2022;**9**:2103414. DOI: 10.1002/advs.202103414

[98] Wohlmuth J, Tekielska D, Cechová J, Baránek M. Interaction of the nanoparticles and plants in selective growth stages—Usual effects and resulting impact on usage perspectives. Plants. 2022;**11**:2405. DOI: 10.3390/plants11182405

[99] Alnaddaf LM, Almuhammady AK, Salem KFM, Alloosh MT, Saleh MM, Al-Khayri JM. Green synthesis of nanoparticles using different plant extracts and their characterizations. In: Al-Khayri JM, Ansari MI, Ak S, editors. Nanobiotechnology: Mitigation of Abiotic Stress in Plants. Cham: Springer; 2021. pp. 165-199. DOI: 10.1007/978-3-030-73606-4_8

[100] Madrid-DelgadoG,Orozco-MirandaM, Cruz-OsorioM,Hernández-RodríguezOA, Rodríguez-Heredia R, RoaHuerta M, et al. Pathways of phosphorus absorption and early signaling between the mycorrhizal fungi and plants. Phyton-International Journal of Experimental Botany. 2021;**90**:1321-1338. DOI: 10.32604/phyton.2021.016174

[101] Ranjan A, Rajput VD, Kumari A, Mandzhieva SS, Sushkova S, Prazdnova EV, et al. Nanobionics in crop production: An emerging approach to modulate plant functionalities. Plants. 2022;**11**:692. DOI: 10.3390/plants11050692

[102] Tarafder C, Daizy M, Alam MM, Ali MR, Islam MJ, Islam R, et al. Formulation of a hybrid nanofertilizer for slow and sustainable release of micronutrients. ACS Omega. 2020;**5**:23960-23966. DOI: 10.1021/acsomega.0c03233

[103] Feregrino-Pérez AA, Magaña-López E, Guzmán C, Esquivel K. A general overview of the benefits and possible negative effects of the nanotechnology in horticulture. Scientia Horticulturae. 2018;**238**:126-137. DOI: 10.1016/j.scienta.2018.03.060

[104] Thavaseelan D, Priyadarshana G. Nanofertilizer use for sustainable agriculture. The Journal of Research Technology and Engineering. 2021;**2**(1):41-59

[105] Solanki P, Bhargava A, Chhipa H, Jain N, Panwar J. Nanofertilizers and their smart delivery system. In: Rai M, Ribeiro C, Mattoso L, Duran N, editors. Nanotechnologies in Food and Agriculture. Cham, Switzerland: Springer; 2015. pp. 81-101. DOI: 10.1007/978-3-319-14024-7_4

[106] Malini S, Raj K, Madhumathy S, El-Hady KM, Islam S, Dutta M, et al. Bioinspired advances in nanomaterials for sustainable agriculture. Journal of Nanomaterials. 2022;**2022**. DOI: 10.1155/2022/8926133

[107] Almuhammady AK, Salem KFM, Alloosh MT, Saleh MM, Alnaddaf LM, Al-Khayri JM. Nanomaterials fundamentals: Classification, synthesis and characterization. In: Al-Khayri JM, Ansari MI, Ak S, editors. Nanobiotechnology: Mitigation of Abiotic Stress in Plants. Cham: Springer; 2021. pp. 77-99. DOI: 10.1007/978-3-030-73606-4_4

[108] Santoro M, Gorelli FA, Bini R, Haines J, vander Lee A. High-pressure synthesis of a polyethylene/zeolite nanocomposite material. Nature Communications. 2013;**4**:1557. DOI: 10.1038/ncomms2564

[109] Sugai T, Yoshida H, Shimada T, Okazaki T, Shinohara H, Bandow S. New synthesis of high-quality doublewalled carbon nanotubes by high-temperature pulsed arc discharge. Nano Letters. 2003:769-773. DOI: 10.1021/nl034183+

[110] Lou XW, Zeng HC. Hydrothermal synthesis of α-MoO3 nanorods via acidification of ammonium heptamolybdate tetrahydrate. Chemistry of Materials. 2002;**14**(11):4781-4789. DOI: 10.1021/cm0206237

[111] Khan NA, Kang IJ, Seok HY, hung SHJ. Facile synthesis of nano-sized metal-organic frameworks, chromium benzenedicarboxylate, MIL-101. Chemical Engineering Journal. 2011:1152-1157. DOI: 10.1016/j.cej.2010.11.098

[112] Jośko I, Oleszczuk P, Skwarek E. Toxicity of combined mixtures of nano particles to plants. Journal of Hazardous Materials. 2017;**331**:200-209. DOI: 10.1016/j.jhazmat.2017.02.028

[113] Samuel MS, Ravikumar M, John JA, Selvarajan E, Patel H, Chander PS, et al. A review on green synthesis of nanoparticles and their diverse biomedical and environmental applications. Catalysts. 2022;**12**:459. DOI: 10.3390/catal12050459

[114] El-Dafrawy SM, Fawzy S, Hassan SM. Preparation of modified nanoparticles of zinc oxide for removal of organic and inorganic pollutant. Trends in Applied Sciences Research. 2017;**12**:1-9

[115] Kanel SR, Manning B, Charlet L, Choi H. Removal of arsenic (III) from groundwater by nanoscale zero-valent iron. Environmental Science & Technology. 2005;**39**(5):1291-1298. DOI: 10.1021/es048991u

[116] Boulkhessaim S, Gacem A, Khan SH, Amari A, Yadav VK, Harharah HN, et al. Emerging trends in the remediation of persistent organic pollutants using nanomaterials and related processes: A review. Nanomaterials. 2022;**12**:2148. DOI: 10.3390/nano12132148

[117] Bizzarri BM, Fanelli A, Botta L, Zippilli C, Cesarini S, Saladino R. Dendrimeric structures in the synthesis of fine chemicals. Materials (Basel). 2021;**14**(18):5318. DOI: 10.3390/ma14185318

[118] Tungittiplakorn W, Lion LW, Cohen C, Kim JY. Engineered polymeric nanoparticles for soil remediation. Environmental Science & Technology. 2004;**38**(5):1605-1610. DOI: 10.1021/es0348997

[119] Ledakowicz S, Paździor K. Recent achievements in dyes removal focused on advanced oxidation processes integrated with biological methods. Molecules. 2021;**26**(4):870. DOI: 10.3390/molecules26040870

[120] Zhang Q, Hou S, Li C. Titanium dioxide-coated zinc oxide nanorods as an efficient photoelectrode in dye-sensitized solar cells. Nanomaterials (Basel). 2020;**10**(8):1598. DOI: 10.3390/nano10081598

[121] Abigail MEA, Samuel SM, Ramalingam C. Addressing the environmental impacts of butachlor and the available remediation strategies: A systematic review. International journal of Environmental Science and Technology. 2015;**12**:4025-4036. DOI: 10.1007/s13762-015-0866-2

[122] Samuel MS, Bhattacharya J, Raj S, Santhanam N, Singh H, Singh NP. Efficient removal of chromium(VI) from aqueous solution using chitosan grafted graphene oxide (CS-GO) nanocomposite. International Journal of Biological Macromolecules. 2019;**121**:285-292. DOI: 10.1016/j.ijbiomac.2018.09.170

[123] Samuel MS, Subramaniyan V, Bhattacharya J, Parthiban C, Chand S, Singh NP. A GO-CS@MOF [Zn(BDC)(DMF)] material for the adsorption

of chromium(VI) ions from aqueous solution. Composites Part B: Engineering. 2018;**152**:116-125. DOI: 10.1016/j.compositesb.2018.06.034

[124] Aslam AA, Aslam AA, Aslam MS, Quazi S. An overview on green synthesis of nanomaterials and their advanced applications in sustainable agriculture. Preprint. 2022:2022020315. DOI: 10.20944/preprints202202.0315.v1

[125] Kumar A, Gupta K, Dixit S, Mishra K, Srivastava S. A review on positive and negative impacts of nanotechnology in agriculture. International journal of Environmental Science and Technology. 2019;**16**(4):2175-2184. DOI: 10.1007/s13762-018-2119-7

[126] Rajput V, Minkina T, Sushkova S, Behal A, Maksimov A, Blicharska E, et al. ZnO and CuO nanoparticles: A threat to soil organisms, plants, and human health. Environmental Geochemistry and Health. 2020;**42**:147-158. DOI: 10.1007/s10653-019-00317-3

[127] Verma KK, Song X-P, Joshi A, Rajput VD, Singh M, Sharma A, et al. Nanofertilizer possibilities for healthy soil, water, and food in future: An overview. Frontiers in Plant Science. 2022;**13**:865048. DOI: 10.3389/fpls.2022.865048

[128] Faisal M, Saquib Q, Alatar AA, Al-Khedhairy AA, Hegazy AK, Musarrat J. Phytotoxic hazards of NiO-nanoparticles in tomato: A study on mechanism of cell death. Journal of Hazardous Materials. 2013;**250**:318-332. DOI: 10.1016/j.jhazmat.2013.01.063

[129] Faisal M, Saquib Q, Alatar AA, Al-Khedhairy AA, Ahmed M, Ansari SM, et al. Cobalt oxide nanoparticles aggravate DNA damage and cell death in eggplant via mitochondrial swelling and NO signaling pathway. Biological Research. 2016;**49**:1-13. DOI: 10.1186/s40659-016-0080-9

[130] Bhardwaj AK, Arya G, Kumar R, Hamed L, Pirasteh-Anosheh H, Jasrotia P, et al. Switching to nanonutrients for sustaining agroecosystems and environment: The challenges and benefits in moving up from ionic to particle feeding. Journal of Nanobiotechnology. 2022;**20**:19. DOI: 10.1186/s12951-021-01177-9

[131] Murata T, Kanao-Koshikawa M, Takamatsu T. Effects of Pb, Cu, Sb, In and Ag contamination on the proliferation of soil bacterial colonies, soil dehydrogenase activity, and phospholipid fatty acid profiles of soil microbial communities. Water, Air, and Soil Pollution. 2005;**164**:103-118. DOI: 10.1007/s11270005-2254-x

[132] Otero-González L, Field JA, Sierra-Alvarez R. Fate and longterm inhibitory impact of ZnO nanoparticles during high-rate anaerobic wastewater treatment. Journal of Environmental Management. 2014;**15**(135):110-117. DOI: 10.1016/j.jenvman.2014.01.025

[133] Punnoose A, Dodge K, Rasmussen JW, Chess J, Wingett D, Anders C. Cytotoxicity of ZnO nanoparticles can be tailored by modifying their surface structure: A green chemistry approach for safer nanomaterials. ACS Sustainable Chemistry & Engineering. 2014;**2**:1666-1673

[134] Zhao J, Wang Z, White JC, Xing B. Graphene in the aquatic environment: Adsorption, dispersion, toxicity and transformation. Environmental Science & Technology. 2014;**48**(17):9995-10009

Chapter 2

Hydraulic Sizing for Watering Green Space Application in Bechar-Algeria

Mebarka Daoudi, Hadja Guedaouria, Youcef Benmoussa, Abdelhak Maazouzi, Karima Seddiki, Fouzia Mehdi, Hanane Maammeri and Asma Belaid

Abstract

Cities with many parks and other green areas are better places to live because they attract more people, improve health and well-being, foster community, and attract tourists and investors. This paper details the detailed process of designing a drip and sprinkler irrigation network to irrigate a 300 m^2 garden in a semiarid region of Algeria. Sprinkler irrigation was then performed using two models of sprinklers: the IPN-5-F, with a radius of 1.5 m, flow of 0.09 m^3/h, and arc of 360°, and the IPN-12H, with a radius of 3 m. The first phase involved hydraulic sizing of the drip irrigation system for varying dripper flows, while the second addressed the use of a drip irrigation system. Drip and sprinkler irrigation have somewhat comparable watering schedules (from one for drip to 2 days for sprinkler), but drip's flow rate is woefully inadequate when compared to sprinklers. Because of the pressure reduction provided by the manifold pipes in a drip irrigation system, it is efficient to install a single pump with a sufficient debt to service both systems.

Keywords: irrigation systems, water requirement, hydraulic sizing, arid region, green spaces

1. Introduction

The presence of green areas in public places helps to improve the quality of air, water, and land by absorbing pollutants, while also working to keep air temperatures within safe ranges, giving shade, stabilizing soil, and reducing harmful carbon dioxide emissions [1]. The negative effects of the current rate of urbanization can be mitigated, in part, by green spaces. They divide up the money fairly between construction, transportation, and basic amenities. Solar energy is an essential alternative clean energy source that can help ensure the long-term viability of our civilization while reducing its environmental impact. With a solar thermal potential of

169,000 TWh/year and a solar photovoltaic potential of 13.9 TWh/year, Algeria is widely regarded as having the greatest solar potential in the entire Mediterranean basin. Whereas in the Algerian Sahara, the average annual sunshine hours are around 3500, which is always over 8 hours per day and can reach up to 12 hours per day in the summer [2, 3]. The focus of this section is on developing a robust solar-powered irrigation system for urban green areas in the semiarid southern region of Algeria (the Bechar region). In this investigation, we focused on two types of watering systems: drip and sprinkler. The following sections outline some of the overarching ideas involved in the development and operation of such systems. **Figure 1** shows how the geometry of the garden was considered during the design of the smart irrigation system.

The following points are the most important criteria to be considered:

- Excessive pressure allows the water to be dispersed into a fine mist, which facilitates the process of evaporation and imbalance in distribution. Weak pressure leads to water flow, poor distribution, and loss of large quantities of it.

- Low pressure causes stagnant water to not only be poorly distributed but also to be wasted in significant quantities.

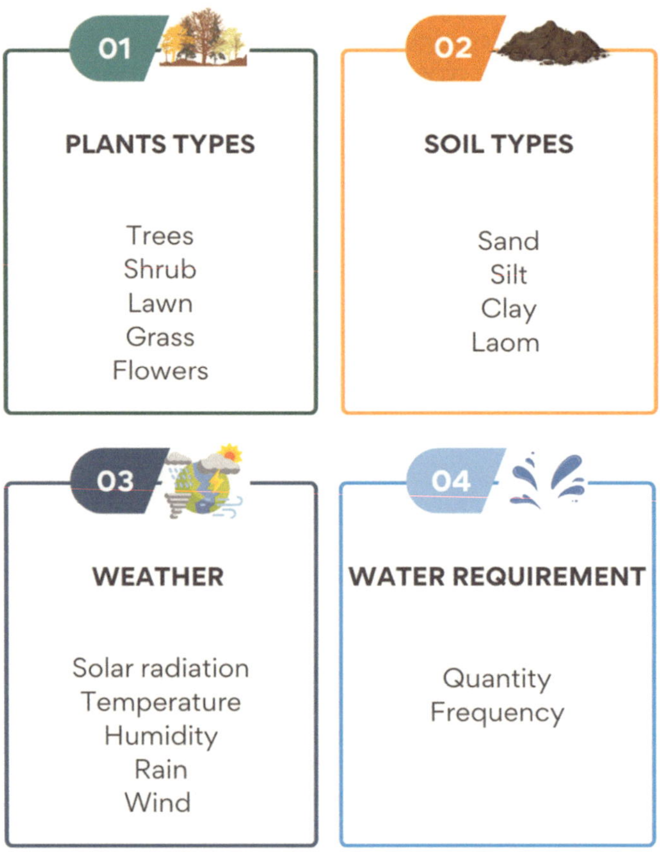

Figure 1.
Requirement parts of the irrigation system.

- The evaporation and dispersal effects of weather and wind on sprinkler irrigation.

- Use flat regions or have them releveled such that they slope only 2%.

It is crucial to confirm the availability of water and its quantity by collecting basic information, such as elevation differences, local climate data, soil type, and crop details (type, growth stage, and pacing).

Determine how much water will be needed at its highest point.

- Decide between a drip and a sprinkler irrigation system.

Do the math on how often you should irrigate and how long it should run.

- Figure out what pipe sizes you will need (supply, main, manifold, and laterals).

Find out how much energy was wasted because of friction.

Determine the sum dynamic head.

Finally, figure out how much energy will be needed for the pumps.

2. Study area

Collect data on elevation, climate, soil, and crops to determine water availability and amount. Determine how much water will be needed at its highest point.

- Decide between a drip and a sprinkler irrigation system.

Do the math on how often you should irrigate and how long it should run.

- Figure out what pipe sizes you will need (supply, main, manifold, and laterals).

Find out how much energy was wasted because of friction.

Determine the sum dynamic head.

Finally, figure out how much energy will be needed for the pumps.

There is a research area of palm trees and flowering shrubbery plants sharing a water demand (**Figure 2**). In the semiarid region of southern Algeria, this green space is part of the University of Bechar in the city of Bechar. It is roughly 300 m^2 in size.

There are 10 palms, evenly spaced at 3 m apart, in the middle section of this landscape, which is watered by a drip irrigation system and has sandy loam soil. Flowers and grass are sprinkle-irrigated in the two nonworking sections. Each palm in the drip irrigation network has two emitters spaced at 1.5 m apart, with each emitter watering an area of 4 m^2 (representing 60 percent of the total wetted area). The following sections summarize the model used to estimate and calculate crop water needed to schedule daily and weekly irrigation and the hydraulic sizing for the two types of irrigation systems: drip and sprinkler systems, respectively presented in the **Figure 2**.

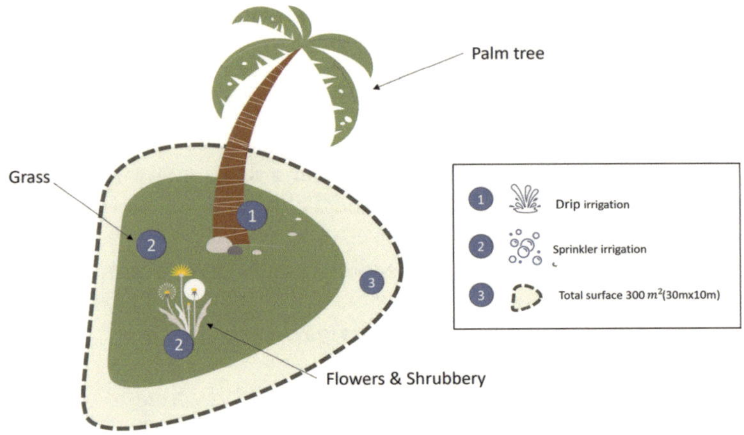

Figure 2.
Part of the study area and the different irrigation systems used.

3. Hydraulic sizing model

3.1 Crop evapotranspiration

Source Plant Since evaporation and transpiration are both influenced by weather, they are climatic parameters. The FAO Penman-Monteith method is the only one suggested for calculating ETo. The technique was chosen because it is physically based, explicitly combines both physiological and aerodynamic characteristics [4] (see **Figure 3**), and provides a close approximation of grass ETo at the region studied.

a. **Evaporation** is the transfer of thermal energy that transforms water from a liquid or solid state into a gaseous state.

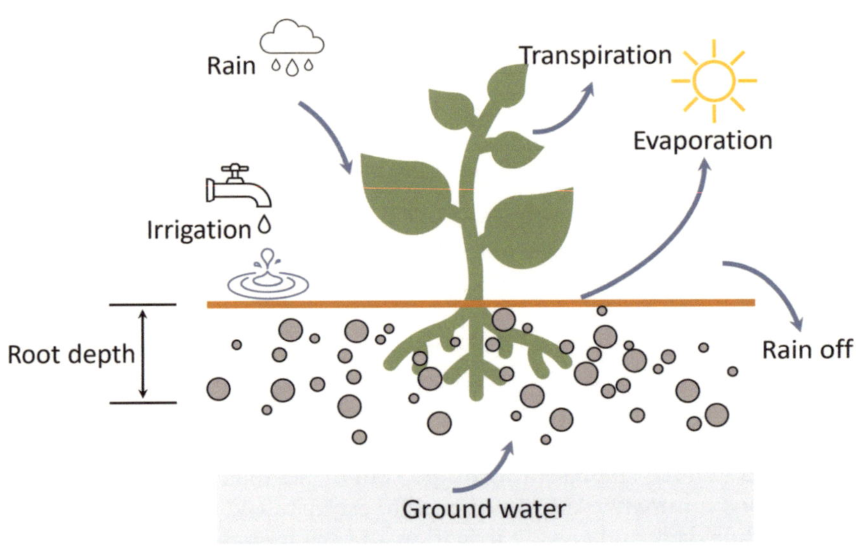

Figure 3.
Physiology and aerodynamic parameters process of a crop.

b. **Transpiration**: The evaporation of water taken up by the crop and used immediately in the construction of plant tissue at a period. Soil evaporation is not a part of this.

c. The term "evapotranspiration" (ET) refers to the total amount of water lost from the evaporation of soil and plants in the area occupied by the crop during its growth. The standard unit of ET is millimeters per day. Evapotranspiration, ET: It is the sum of the amount of water transpired by plants during the growth process and the amount that is evaporated from soil and vegetation in the domain occupied by the growing crop. ET is normally expressed in mm/day.

This refers to crop ET under standard conditions, that is, *ET* from disease-free, well-fertilized crops, grown in extensive fields, under optimum soil water conditions. ET_c can be derived from ETo using the equation [4]:

$$ET_c = ET_0.k_c \tag{1}$$

where:
ET_0: Reference evapotranspiration.
k_c: Crop factor.
Crop evapotranspiration under nonstandard conditions, as mentioned above, is called ETc (adjusted). This refers to the growth of crops under no optimal conditions [4].
For more specification case of drip irrigation system, crop evapotranspiration stated in Eq. (1) can be given by [5]:

$$ET_{c-drip} = ET_c.k_r \tag{2}$$

k_r : Ground cover reduction factor.

3.2 Water application net depth

Water should be delivered at a certain depth during irrigation to make up for what the crop loses *via* evaporation and transpiration. The following parameters are needed for the net depth of water application computation [6]:

- Moisture in the ground (Field capacity predicted wetness)

- Permissible loss of soil moisture (P).

The RZD is the effective depth of the crop's root zone.
It is important to survey and test the soil to learn its field capacity (FC) and permanent wilting point (PWP). When the soil texture is understood, it is possible to make estimations without the equipment and time by looking up numbers in the literature, ideally from the area. However, there is a discrepancy between the reported data on the moisture content of various soil types.
The maximum net depth to be applied per irrigation can be calculated using the following equation [6]:

$$d_{net} = (F_C - PWP).RDZ.P \tag{3}$$

where:

d_{net}: Readily available moisture or net depth of water application per irrigation for the selected crop (mm).

F_C: Soil moisture at field capacity (mm/m).

PWP: Soil moisture at the permanent wilting point (mm/m).

RDZ: Depth of soil that the roots exploit effectively (m).

P: Allowable portion of moisture permitted for depletion by the crop before the next irrigation

To express the depth of water in terms of the volume, the area proposed for irrigation must be multiplied by the depth |6]:

$$Volume\ of\ water\ to\ be\ applied\ \left(m^3\right) = 10.A.d_{net} \qquad (4)$$

where:

A: Proposed area for irrigation (ha).

d_{net}: Depth of water application (mm).

3.3 Irrigation efficiency

How these functions (I_f) interact regarding uniformity and losses determines the irrigation efficiency [7].

$$I_f = \frac{d_{net}}{ET_c} \qquad (5)$$

There are some factors that affect irrigation efficiency which are as follows:

- *System design*: Along with maintenance requirements and site conditions, these are fundamental factors that influence the overall efficiency of irrigation systems.

- *Soils*: High irrigation efficiencies are easier to achieve on heavier and deeper soils, while low efficiencies are more common for sites on shallow and free-draining soils.

- *Sprinkler packages*: The right sprinkler package that is well maintained can reduce losses to evaporation and improve the distribution uniformity and application efficiency.

- *Pump flow rate*: High flow rates for flood irrigation systems can reduce the time required to water each bay and hence use less water per irrigation event.

- *Irrigation scheduling*: Applying the right amount at the right time to match crop water requirements and soil moisture holding capacity can cause less deep drainage and improved efficiency.

- *Application depth*: If the average application depth is greater than the soil moisture holding capacity, it will cause excessive deep drainage per irrigation event and, therefore, contribute to low overall system efficiency [8].

The factors to improve irrigation efficiency are as follows:

- Water losses can be decreased to increase irrigation effectiveness. The following three key areas could use improvement.
- Transmission system: Alternatives for channel lining or treatment can lower channel system losses. To prevent excessive losses, piped systems need to be properly maintained.
- Application system: Achieving high irrigation efficiency requires effective irrigation system design and maintenance.
- In managing irrigation systems, deep drainage losses are reduced when the amount of water is applied at the time [8].

3.4 Irrigation requirement

FAO 1984 defines the net irrigation requirements (IR_n) as the depth or volume of water required for normal crop production over the whole cropped area, excluding contribution from other sources. The following equation is used to calculate the amount of irrigation requirement [9]:

$$IR_n = (ET_c.k_r) - R + LR \tag{6}$$

k_r: Ground cover reduction factor
R: Water received by plant from other sources than irrigation (for example rainfall,)
LR : Amount of water required for leaching of salts

Gross irrigation need is the term used to describe the total amount of water applied through irrigation. The net irrigation requirement plus water application losses and other losses make up this amount. Depending on the requirements, the gross irrigation need can be calculated for a field, a farm, an outlet command area, or an irrigation project by considering crop losses at different phases of growth [10].

$$IR_g = \frac{ET_c.k_r}{E_a} - R + LR = \frac{IR_n}{E_a} \tag{7}$$

where:
IR_g: Gross irrigation requirement (mm/day).
E_a: Field application efficiency of drip irrigation system

According to Rain Bird International (1980), the following efficiencies should be used when the surface area wetted by one emitter does not exceed 60 cm in diameter [9]: Hot dry climate: 0.85, Moderate climate: 0.90, Humid climate: 0.95.

3.5 Emitter sizing

The number of emitters per plant is established as follows [9]:

$$N_p = \frac{area\ per\ plant.P_w}{A_w} \tag{8}$$

P_w: Percentage wetted area.
A_w: Area wetted by one emitter.

The diameter of the wetted area can be calculated using this equation:

$$A_w = \frac{\pi.D^2}{4} \tag{9}$$

The percentage of wetted area P_w should be checked to see if it is still within the recommended limit. This can be done using the equation [8]:

$$P_w = \frac{100.N_p.S_e.W}{S_p.S_r} \tag{10}$$

where:
S_e: The distance between the emitters, which is the emitter spacing.
W: Wetted width or width of width strip along lateral with emitters.
S_p: Distance between the plant within a row.
S_r: Distance between plant rows or rows spacing.
The relation between S_e, S_p, and N_p given the spacing, which can be calculated by [9]:

$$S_e = \frac{S_p}{N_p} \tag{11}$$

The duration of irrigation or length of operation time (T_a) at peak demand can be established by the equation below [9]:

$$T_a = \frac{IR_g}{N_p.q} \tag{12}$$

T_a: Duration of irrigation per day
q : Emitter discharge

3.6 Sprinkler sizing

3.6.1 Set time

Set time is the time each set of sprinklers should operate at the same position to deliver the gross irrigation depth, which can be calculated by [11]:

$$T_s = \frac{d_{gross}}{P_r} \tag{13}$$

where:
T_s: Set time
P_r: Sprinkler precipitation rate
d_{gross} : Gross depth of water
The net depth of irrigation, divided by the agricultural irrigation efficiency, yields the gross depth of water application (E). It should be emphasized that possible water losses from pipe leaks are factored into the effectiveness of farm irrigation. The gross depth of water can be calculated using the equation shown below [11]:

$$d_{gross} = \frac{d_{net}}{E} \tag{14}$$

E : Farm irrigation efficiency

3.6.2 Sprinkler system capacity

Sprinkler system capacity is one of the predetermining parameters in the evaluation of a sprinkler irrigation system. It depends on the system application rate. The sprinkler discharge test is carried out and compared with the calculated value using equation [11]:

$$Q = N_c.N_s.Q_s \tag{15}$$

Q: The system capacity
N_c: The number of laterals operating per shift while
N_s: The number of sprinklers per lateral
Q_s: The sprinkler discharges

The actual selection of the sprinkler is based on design information furnished by the manufacturers of the equipment. The choice depends mainly on the diameter of coverage required, pressure available, and sprinkler discharge. The sprinkler discharge may be determined by the following formula [11]:

$$Q_s = \frac{S_I.S_m.I}{360} \tag{16}$$

S_l: Spacing of sprinklers along the laterals
S_m: Spacing of laterals along the main
I : Optimum application rate

The number of sprinklers calculated by [12]:

$$N_s = \frac{field\ width}{S_l} \tag{17}$$

- The goal of irrigation is to supply enough water to keep the crop's root zone at the optimal level of soil moisture. Irrigation work must be completed at a fair labor, power, and material cost.

Below, I summarize the benefits of sprinkler irrigation over surface irrigation:

- Most soils with an infiltration rate under 4 cm/h can use it.
- Appropriate for practically all crops.
- It is appropriate to use uneven ground.
- Irrigation water is a cost-effective way to apply fertilizer, herbicides, and fungicides.
- Can be used to cool crops in the summer and protect against winter frost.

- Bunds and supply channels are not necessary.
- Saving money on labor and water.
- Enables the transportation of farm equipment.
- Increased crop yields and healthy growth.
- Less disease and pest infestation.
- Sprinkler irrigation restrictions.
- Sprinkler patterns become distorted by wind, leading to uneven distribution.
- Soft fruit that is ripening has to be shielded from spray damage.
- Higher upfront costs.
- High power needs because it operates at 5 to 100 m of water head.
- Not appropriate for fine-textured, slow-infiltrating soil.
- After irrigation, the movement of portable pipes in some soils could be problematic.

3.7 Pipes sizing

After computing the friction losses in the supply line, mainline, manifold, and laterals for a more in-depth check reference [13, 14], it is important to first calculate the permissible pressure fluctuation that will offer emission uniformity close to the required value.

3.8 Total dynamic head

The total dynamic head (TDH) of a pump is the sum of the total static head, pressure head, friction head, and velocity head. An explanation of these terms follows and is shown graphically in **Figure 4** [15].

3.8.1 Total static head

The vertical distance that the pump must raise the water is known as the total static head. When pumping from a well, the vertical lift of the water from the ground surface to the discharge point is added to the distance from the well's pumping water level to the ground surface. It would be the total vertical distance from the water surface to the discharge point when pumping from an open water surface [15].

3.8.2 Pressure head

Pressure is necessary for sprinkler and drip irrigation systems to function. To distribute the water effectively, center pivot systems require a specific pressure at the pivot point. By multiplying by 2.31, the pressure head at every location, where a pressure gauge is installed, can be translated from PSI to feet of head.

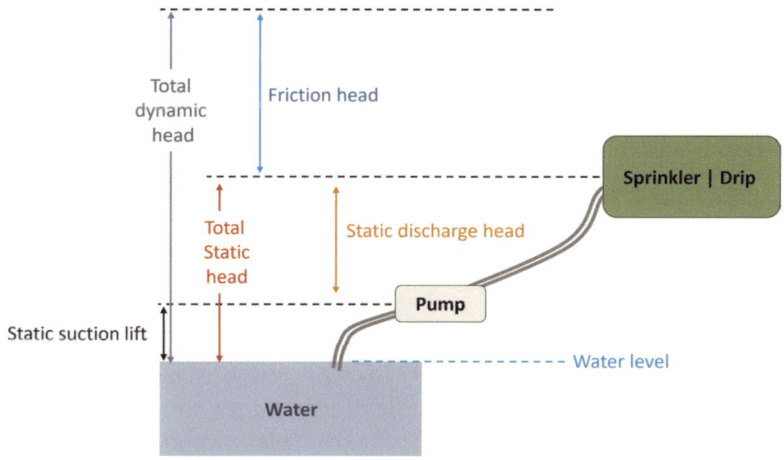

Figure 4.
Total dynamic head [15].

3.8.3 Friction head

The energy or pressure reduction brought on by friction when water moves through pipe networks is known as the friction head. The amount of friction loss is significantly influenced by the water's velocity. When water flows through straight pipe sections, fittings, or valves, around corners, or where pipes expand or contract in size, frictional head loss occurs. These losses' values can be computed or got from tables of friction loss. The total of all friction losses determines the friction head of a piping system [15].

3.8.4 Velocity head

The energy of the water is known as velocity head because of its velocity. When calculating irrigation system losses, this smaller amount of energy is typically negligible [15].

3.9 Pumping power requirement

The choice of the pump and power plant is the next matter to be addressed in a design. A pump that should deliver the head and flow at the best efficiency, as well as an electric motor to drive the pump, should be chosen from the manufacturer's charts. The criterion for choosing a pump is whether its Net Positive Suction Head Required (NPSHR) is greater than it's Net Positive Suction Head Available (NPSHA). The basic formula for calculating power requirements is given below [6]:

$$Power\ requirement\ in\ BHP = \frac{Q.THD}{273.E_p} \tag{18}$$

$$Power\ requirement\ in\ KW = \frac{Q.THD}{360.E_p} \tag{19}$$

where:
KW or BHP: Energy transferred from the pump to the water
Q: Discharge
TDH: Total dynamic head
E_p: The pump efficiency (%) from the pump performance chart.
It should be noted that this formula expresses the actual power required at the pump

4. Results and discussion

First, the climatic data collection for Bechar-Algeria semiarid region during the corn period, which takes place in the middle of July as; the minimum and maximum temperature ($T_{min} = 25^{\circ}$ and $T_{max} = 40^{\circ}$), the day solar radiation ($6944.4\ W/m^2/day$), the sunshine hours (10.4 hours), the humidity (33%), and the wind speed (156 Km/j) needed for running the calculation, after the collection of the principal data for hydraulic sizing is necessary. Using the data shown in **Table 1**, the results of designing irrigation networks are presented below.

The characteristics of the irrigation system can be simply determined if adequate parameter values have been estimated for each type of crop and area. Each palm will have two emitters in the drip irrigation network, spaced 1.5 m apart. About 4 m^2 is wetted by one emitter, with 60% of that area being wet. It is crucial to compare several dripper laws in order to select the best dripper flow. The results of sizing for variations in dripper flow from 4 to 23 l/h are shown in **Table 2**.

Table 2 shows that when pipe flow and diameters increase, irrigation hours drop and pipe friction loss (at laterals and manifolds) increases. The total dynamic head rises because of friction loss. As a result, the energy needed to run the solar pump increased from 15 to 88 W. Because irrigation frequency depends on how much water the crop needs, it was constant across many drippers. The dripper of 15 l/h for pump

Crop	RDZ (mm)	ETP (mm/day)	Efficiency (%)	FC-PWP (mm/m)
Palm tree	0.7	7.4	55	90
Grass	0.5	8	55	90

Table 1.
Principal data for hydraulic sizing [5, 6, 9, 16].

Dripper flow (l/h)	4	8	15	23
Irrigation frequency (day)	1	1	1	1
Irrigation hours (h)	8	4	2	1.44
Total lost friction in pipes (m)	0.35	0.5	0.9	2.03
Pump flow (m^3/h)	0.16	0.32	0.60	0.92
Total dynamic head (m)	12.50	17.50	18.93	19.02
Power requirement (w)	15	30	57	88

Table 2.
Results of sizing drip irrigation network with different dripper flow.

Sprinkler code	IPN-5-F	IPN-12H
Precipitation (mm/h)	33.5	22
Number of sprinklers	17	7
Pump flow (m^3/h)	3	1.5
Power requirement (w)	263	124
Irrigation hours	1	1.5
Irrigation frequency (day)	2	2

Table 3.
Results of sizing irrigation network with two models of sprinkler.

debit is approximately 0.6 m^{-3}./h is the best option in terms of operation times and power requirements.

Two sprinkler types have been employed in the sprinkler irrigation system: the IPN-5-F, which has a radius of 1.5 m, a flow of 0.09 m^3/h, and an arc of 360°, and the IPN-12H, which has a radius of 3 m, a flow of 0.1 m^3/h, and an arc of 180°. **Table 3** compiles the findings from sizing.

In terms of the number of sprinklers, pump flow, and power needed, the sprinkler IPN-12H is a better option than the sprinkler IPN-5F. For the best design, we installed sprinklers IPN-12Q with a 90° arc in the corners. For additional information on these sprinkler types, check the reference [9]. The sprinkler spacing is around 4.2 m because of the sprinkler head spacing percentage, which is approximately 60% (for the desert region) [17, 18].

For the two types of irrigation systems—drippers and sprinklers—the irrigation frequency values range from 1 to 2 days, respectively. The pump debit in the drip irrigation system (0.6 m^3/h) is minimal compared to the pump debit in the sprinkler irrigation system (1.5 m^3/h). In addition, the irrigation hours in the two networks are different: the drip irrigation network is closed for two hours, while the sprinkler irrigation network is closed for 1.5 hours. These factors make it appropriate to install only one pump, which has the greater debit for the two networks and the drip irrigation system; manifold pipes must use decreased pressure.

There are some situations that could influence and interpret how much power is needed to run the pump and the entire dynamic head. Three situations are examined here, and **Figure 5** shows the outcomes:

a. *Case 1*: Variation of distance between the water source and the green space from 10 to 100 m, see **Figure 5(a)**.

b. *Case 2*: Variation of difference in elevation between the water level and the highest point on the land from 1 to 10 m, see **Figure 5(b)**.

c. *Case 3*: Variation of suction lift from 1 to 5 m, see **Figure 5(c)**.

In **Figure 5(a)–(c)**, the power requirement (*Pr*) increases relatively from 126 to 200 W because the total dynamic head (*TDH*) boost slightly from 18 to 27 m. In **Figure 5(a)**, whatever the length of supply line rise, the total friction losses do not exceed the allowable pressure variation, which equals 1.5 m in our case, this is achieved by changing of pipes section. In **Figure 5(b)**, a slope of two percent or less is

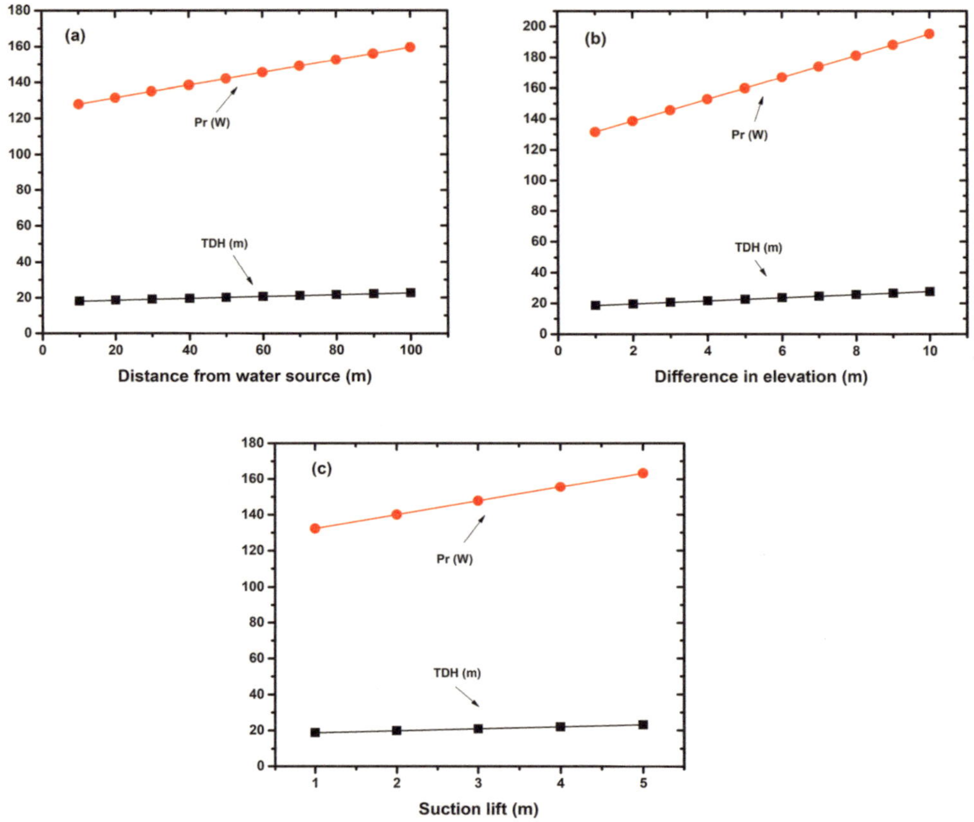

Figure 5.
Different variations of (a) distance between the water source and the green space, (b) difference in elevation between the water level and the highest point on the land, and (c) suction lift.

excellent for irrigation systems; any slope more than this causes pressure changes, which require more power to maintain the required flow rate. Use of pressure-compensated drippers is strongly advised in this circumstance [11].

The third scenario (**Figure 5(c)**) shows a slight increase in TDH and Pr that only slightly affects the suction lift, which for a surface pump does not exceed 5 m.

The performance of seasonal crop water needs, ETc, I-f, and T at peak demand for the two types of irrigation networks should be studied, it is concluded. **Table 4** presents the outcomes.

Month	ET_C (mm/day)	I_f (day)		T (h)	
		Drip (q = 15 l/h)	Sprinkler (R = 3 m)	Drip (q = 15 l/h)	Sprinkler (R = 3 m)
Winter	4	3	5	1	0.5 h
Spring	6	2	4	2	0.7 h
Summer	8	1	2	3	01 h
Fall	5	2	4	2	0.7 h

Table 4.
Performance of seasonal crop water needs for irrigation system.

The findings are shown in **Table 4**, which includes the variation in irrigation time (T) and frequency (I-f) for the crop evapotranspiration (ET-C) values for the season. The minimum and maximum values were attained during the corn period (period of summer), and then in the fall, as shown in the graph below. The changing of seasons is caused by I-f. and T. Irrigation becomes rarer during winter, then rises again in summer. Longer irrigation times and fewer cycles are caused by increased crop water demand in summer because of higher temperature. Both drip and sprinkler irrigation networks are linked to the same irrigation system. It is suggested to adjust the irrigation value duration. This investigation will be released later.

5. Conclusion

The application, effectiveness, and financial return of irrigation in the industrial process can all be improved by improving the design of irrigation systems. Design requirements should be based on an understanding of irrigation, hydraulics, economics, energy, and environmental factors. The principal goal of this study is to arm the general people with the information and resources required to achieve the effective use of water for outdoor irrigation, as well as those who are to water public green spaces. A network of permanent pipes joined to emitters and sprinklers that are installed and planned to use the water from a particular natural area is an irrigation system. Tools installed on a tube and used under pressure to drain water as a drip or nozzle are emitters and sprinklers, respectively. A system that is established and intended to lower the capacity for producing water is an efficient irrigation system.

Acknowledgements

This work is supported by the General Direction of Scientific Research and Technology (DGRSDT) of the Algerian Ministry of Higher Education and Research (social-economical project 2021).

Nomenclature

ET_{crop}	Crop evapotranspiration (mm/day)
ET_0	Reference evapotranspiration (mm/day)
$ET_{CROP-drip}$	Crop evapotranspiration for drip irrigation (mm/day)
k_c	Crop factor (0.9)
k_r	Ground cover reduction factor (0.8)
d_{net}	Net depth (m)
$F_C - PWP$	Moisture (90 mm/m)
RDZ	0.5m | 0.7m
P	Allowable portion of moisture for drip | sprinkler (20% | 50%)
A	Irrigation area (ha)
d	Water application depth (mm)
IR_n	Net irrigation requirement (mm)
IR_g	Gross irrigation requirement (mm)

R	Water received by plants from sources other than irrigation (mm/day)
LR	Amount of water required for leaching of salts (mm/day)
E_a	Field application efficiency of drip irrigation system (0.85)
d_{gross}	Gross depth of water application (m)
E_S	Farm irrigation efficiency of sprinkler irrigation system (0.65)
q	Dripper discharge (l/h)
N_D	Number of drippers per plant (---)
P_w	Percentage of wetted area (60%)
A_w	Area wetted by one emitter (m^2)
S_D	Drippers' spacing (m)
S_p	Plant spacing (m)
q	Emitter discharge (l/h)
T_s	Set time (h)
P_r	Sprinkler precipitation rate (mm/h)
d_{gross}	Gross depth of water (mm)
T_a	Duration of irrigation per day (h)
N_S	Number of sprinklers (---)
Q	The system capacity (m^3/h)
N_c	Number of laterals operating per shift while (---)
N_s	Number of sprinklers per lateral (---)
Q_s	Sprinkler discharge (---)
S_l	Spacing of sprinklers along the laterals (m)
S_m	Spacing of laterals along the main (m)
I	Optimum application rate (cm/h)
E_p	Pump efficiency performance chart (%)
TDH	Total dynamic head (m)
KW or BHP	Energy transferred from the pump to the water (W)
Q	Discharge (m^3/h)

Author details

Mebarka Daoudi[1]*, Hadja Guedaouria[1], Youcef Benmoussa[1], Abdelhak Maazouzi[2], Karima Seddiki[2], Fouzia Mehdi[1], Hanane Maammeri[1] and Asma Belaid[1]

1 LPDS Laboratory of Semiconductor Devices Physics, University of Bechar, Bechar, Algeria

2 LCSE Laboratory of Chemistry and Environmental Sciences, University of Bechar, Bechar, Algeria

*Address all correspondence to: daoudi.mebarka@univ-bechar.dz

References

[1] World Health Organization. Urban Green Space Intervention and Health. European Environment and Health Process; 2017

[2] Chancelier L, Laurent E. l'électricité photovoltaïque. Cooperation Francaise, Ademe; 1995

[3] Available from: https://www.aps.dz/ar/regions/109002-2021-06-26-12-18-25

[4] Available from: https://www.dpg polytechnic.com/downloads/files/n5acb055ce5b1c.pdf

[5] Doorenbos J, Pritt WO. Crop water requirements crop water requirements, no. 24. 1977

[6] Sprinkler Irrigation System Planning, Design, Operation and Maintenance Volume III Module8. Harare: FAO, Sub-Regional Office for East and Southern Africa (SAFR); 2001

[7] United States Department of Agriculture. Chapter 2 irrigation water requirements (210-vi-NEH, September 1993) 2-1. In: Irrigation Water Use Bookmarks and Buttons to Navigate. Issued September 1993

[8] Harvey A, Palamountain R, Zerk M, Heffernan J, Newson D, Powell T, "Irrigation glove boxguide". 2016

[9] D. Monitoring, I. Agriculture, F. Participation, and V.Iv. Irrigation Manual Module 9 Lpcalized Irrigation Systems Planning, Design, Operation and Maintenance. Vol. IV. 2022

[10] Available from: https://ebrary.net/121143/environment/gross_irrigation_requirement

[11] Suray A. Rasheed, Aqeel Al-Adil, Sprinkler Irrigation Systems and Water Saving, a Case Study from South of Iraq, 2nd Conference on Environment and Sustainable Development 28–29-Oct-2015, Water Resource Ministary, Baghdad

[12] Shiva Shankar M, Ramanjaneyulu AV, Neelima TL, Das A. In: Rajkhowa DJ, Das A, Ngachan SV, Sikka AK, Lyngdoh M, editors. Sprinkler Irrigation – An Asset in Water Scarce and Undulating Areas. Integrated Soil and Water Resource Management for Livelihood and Environmental Security. Meghalaya, India: ICAR Research Complex for NEH Region; 2015. pp. 259-283

[13] Proceedings of Drip and Sprinkler Workshop. Jalgaon (M.S.), India; 1994

[14] Jack K, Ron B. Sprinkler and Trickle Irrigation. Caldwell, New Jersey: The Blackburn Press; 2000. ISBN: 1-930665-19-9

[15] Scherer TF. Irrigation Water Pumps-Publication. Vol. 1057. NDSUF; August 2017. pp. 1-12

[16] Sudarshan S. Affordable Micro Irrigation Technology. Delhi, India: International Development Enterprises; 2000

[17] Mthieu C, Audoye P, Chossat J. Bases techniques de l'irrigation par aspersion. LAVOISIER; 2007

[18] Hadja Guedaouria et al. 2022 IOP Conf. Ser.: Earth Environ. Sci. 973 012013

Chapter 3

Perspective Chapter: The Impact of Human Activities on Wetlands' Provisioning and Cultural Services in Epworth, Zimbabwe

Tatenda Musasa, Kudakwashe C.R. Muringaniza and Emmanuel Dakarai Mhlanga

Abstract

The study assessed the impacts of human activities on wetlands' provisioning and cultural services in Epworth's Ward 6 (Overspill), Zimbabwe. The researcher adopted a case study approach which implemented both the qualitative and quantitative approach. Semistructured interviews, questionnaires and field observations were conducted to gather primary data that was synthesised with collected secondary data. Data obtained from the field was analysed through Statistical Package for Social Scientists (SPSS) software version 20.0 and Microsoft excel. Findings show that the majority of households (31.6%) were motivated to utilise wetlands due to the availability of market for agricultural products. Wetland use was influenced by population increase, availability of rainfall and donor funding for sustainable food security projects. The dominant wetland use is agriculture as indicated by 35.1% of respondents. Essential services of the wetland include supply of water for agriculture (26.3%), livestock grazing, spiritual enrichment, cultural heritage, ecotourism etc. Agriculture, urbanisation, dumping of waste and livestock grazing negatively impacted the wetland. The study advocates for support of environmental conservation initiatives from key stakeholders. It also recommends that relevant authorities come up with an integrated land use planning approach that will go a long way in addressing issues of land degradation and fragmentation.

Keywords: wetlands, human activities, provisioning services, cultural services, Epworth

1. Introduction

Wetlands are defined as areas of marsh, fen, peat land, or water, whether natural or manmade, permanent or temporary, with water that is static or flowing, fresh, brackish, or salt, including sea water with a depth of less than 6 m at low tide [1]. Wetlands provide benefits to humans and the environment which can be categorised as provisioning, cultural and supporting [2, 3]. Provisioning services are those benefits provided or produced by wetland ecosystem for example food, fuel, water and fibre or genetic resources whereas cultural services are non-material benefits from wetland ecosystems [3–5]. Cultural services include spiritual enrichment, ecotourism, recreation, aesthetics and cultural heritage while supporting services are factors which are necessary for producing ecosystem services for example hydrological cycle, soil formation and nutrient cycling [6].

Wetlands play an essential part in people's life as supplies of water for residential and irrigation needs, important fishing regions, plant harvesting for roofing and crafts manufacture, food production, and valuable grazing land [7, 8]. Wetlands have provided substantial opportunities for tourism and recreation on a global scale, providing financial benefits to governments, tourism industry, and local residents, with the proceeds being used to finance their protection [9]. Human activities are posing a threat to wetlands all over the world. In 1971, nations assembled in Iran for the Ramsar Convention to address concerns about wetlands and their deterioration [9]. According to Mandishona and Knight [10], the poor directly rely on natural resources for their livelihood, and high poverty levels in most emerging countries have accelerated the rate of wetlands destruction. Kimani [11] further expressed that in developed countries, the consequences of wetlands loss and degradation are frequently mitigated with costly artificial constructions, such as major flood protection schemes or water purification plants, but this is not the case in developing countries, where mitigation measures may take too long to implement, owing to financial and technical constraints. More than half of the world's population lives in cities, with that number expected to climb to 75% by 2050 [12].

Wetlands in India for example are under threat from overpopulation as well as a variety of human activities such as the construction of houses and roads, agriculture, and overfishing [11, 12]. Kumar [13], opines that agriculture and urbanisation in India have taken over the wetland area due to population growth which has led to expansion of human activities into wetlands hence affecting wetlands provisioning, cultural and supporting services. Ahidur [7], concluded that most people in India are illiterate and live in an impoverished location, they primarily participate in primary occupations such as fishing.

Africa is one of the most urbanised continents, yet this phenomenon has not kept pace with the continent's economic, social, and political progress [8, 14]. In Africa, urbanisation has been increasing by the rate of approximately 3.9% annually. According to United Nations Population Fund, it is estimated that the population of Africa will be doubled by the period of 2030. High human population growth rates and economic development in Kenya have also resulted in wetlands, forests, ecosystems, and air quality degradation [5]. Sedimentation, livestock wastes, and hazardous wastes, pose a severe threat to lakes. These are washed away from the earth and wind up in rivers, streams, lakes, and seas, causing eutrophication and altering food chains. Most wetlands in Kenya are being impacted by an expanding human population, which means more land is being opened up for cultivation, necessitating increased use of pesticides and other agro-chemicals, which are then washed down through surface run-off, resulting in wetlands pollution [15].

Wetlands in Zimbabwe are under pressure from human activities. Over the last few decades, Zimbabwe has seen a steady loss of wetlands, resulting in major societal losses [6]. In Zimbabwe since 1980 after the country gained independence, the population of the country's major cities is increasing by the rate of 5% per annum [16]. According to Marambanyika and Beckedahl [5], both the colonial and independence governments' socio-economic policies and political expediency had a considerable impact on the morphology (form, structure, and population distribution) of Zimbabwe's towns and cities, affecting wetlands ecological services. Due to increased demand for services, Epworth's rate of urbanisation and population growth has mostly outpaced existing resources [17]. Based on this background, the goal of this study was to assess the effects of human activities on wetlands' provisioning and cultural services in Epworth. This was achieved through the following objectives to; (1) determine the major land use activities in wetlands located in Epworth, (2) establish the factors influencing wetland use as perceived by the households, (3) assess the current range of wetland provisioning and cultural services obtained by the local community, (4) examine the impact of human activities on wetland provision and cultural services in Epworth, Harare.

2. Materials and methods

2.1 Location of study area

The study area is Ward 6 (Overspill) of Epworth District which is situated in Zimbabwe's Harare Province and is located about 12 km from Harare city centre (**Figure 1**). It is located on the north-eastern side of Epworth district and is the second biggest ward in the district between latitudes 17.873°S and 17.899°S and longitudes 31.156°E and 31.203°E. Overspill is one of the most densely populated wards in Epworth District [15]. There is a stream flowing towards Manyame River through the midsection of Overspill which runs seasonally, and in the summer, completely drowns wetlands. Overspill as part of the Southern Africa Highveld, is characterised by gradually undulating ground punctuated by granite outcroppings and magnificent balancing rocks that are popular with tourists and mildly rolling terrain in the upland parts [18, 19]. The soils of the area are primarily silt to clay, characterised prominently by riparian vegetation in the wetlands dominated by reeds. The suburb has been selected as the study area because it consists of the largest informal settlement across urban districts in Harare. In 1980, Epworth had a population of 20,000, in 2002 it had an estimated population of 123,250 and in 2012 it had 167,462 [16]. It is largely dominated by high density residential areas. Ward 6 (Overspill) has a population of 25,597 people.

2.2 Data collection and analysis

Field data were collected between October and November 2021 to solicit data on human impacts on wetlands. The study adopted a case study research design which allowed collection of the data in a realistic setting and increased the researcher's analytical power as well as ability to learn more about the social issues. The quantitative and qualitative approaches were utilised in the study to explore the impacts of human activities on wetlands provisioning and cultural services in Overspill, Epworth. These techniques utilised semi structured questionnaires, interviews, observations and secondary data from principal land uses on wetlands records. The researchers selected representatives from Overspill (Ward 6) households to respond to the questionnaire as they formed the majority of people who are resident near wetlands and hence possess diverse knowledge on wetland utilisation in the area. Households were also targeted to identify the drivers of the activities that are being done on wetlands and

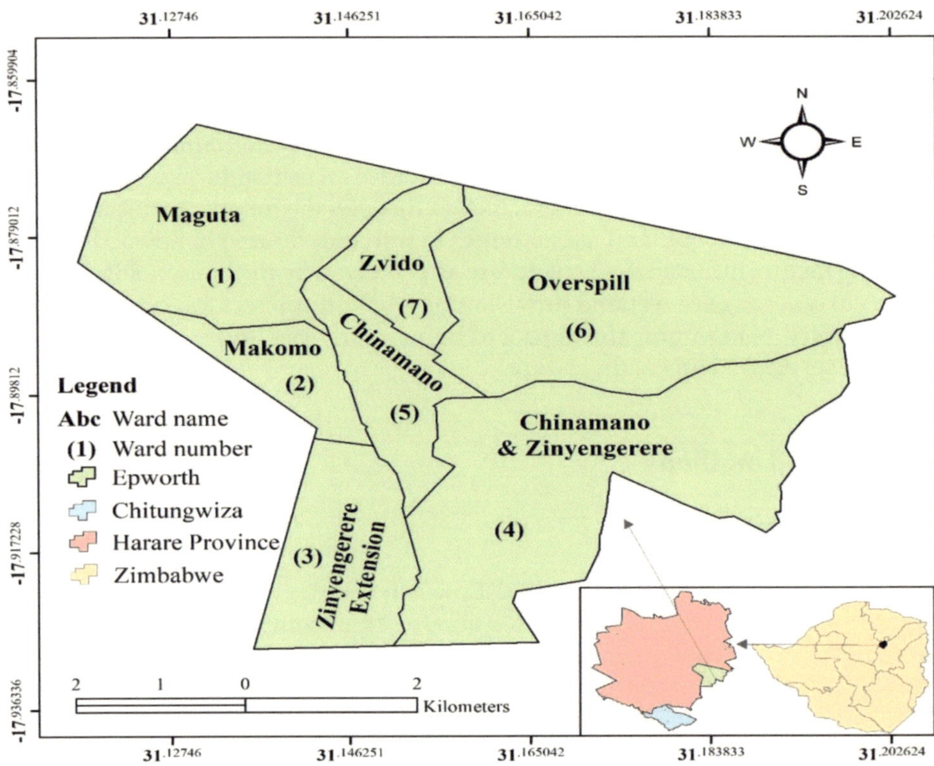

Figure 1.
Ward 6 of Epworth District, Harare Province.

also the impacts of these activities on this ecosystem. Targeted key informants for the interviews included Epworth Local Board employees, Environmental Management Agency (EMA) Officer, Ward 6 councillor and Harare City Council Urban Planning Department as well as Housing Department officials. The officials from Harare City Council's Urban Planning Department and Housing Department were targeted because they are in charge of administering land within the study area as well as issuing licences. Officials from the Departments of Housing and Engineering were targeted because they have an environmental section in charge of managing all natural resources, including wetlands hence it is their responsibility to safeguard and monitor the environment. In line with the study's goal, the EMA provided useful information on human activities that affect wetlands as it is the major stakeholder in responsible for wetland management and protection. The councillors were chosen because they represent legislators of the by-laws that govern local wetland use.

Clustered random sampling was utilised to pick target respondents who lived within 1000 m of a wetland within the study area. The targeted households were regarded as part of a heterogeneous community in terms of their benefits from wetlands due to their location. The study adopted the use of random systematic sampling on the households located within 1000 m of concerned wetlands in Overspill (Ward 6). A ground census was carried out in the field to identify the number of households located within a 1000 m radius of wetlands in Overspill, Ward 6 of Epworth District and 300 households were identified in this process. A total of 60 households (20%) was determined as the sample size of questionnaire respondents. A random household was chosen as the starting point from which every fifth household was selected for the study amounting to 60 households. The sampling criteria gave each household an equal chance of selection and reduced any form of bias. The key informants were purposively sampled which allowed the selection of professional interview respondents who possessed attributes or experience requisite in the understanding of the research's objectives.

The researcher administered face to face questionnaires in order to get in-depth responses from respondents on their perceptions about wetlands. Questionnaires allowed the researcher to capture wider and varying views of participants with regards to wetlands provisioning and cultural services. The questionnaire also allowed respondents to respond freely and better express their positions and opinions towards the effects of human activities on wetland and what they thought could be remedial on the human activities on wetlands. The surveys were self-administered to assure the formulation of complex questions, especially for illiterate people, and to save money. Since it was during COVID-19 era, the researcher took the survey under COVID-19 regulations that is practising social distance, wearing of masks when questioning questions to the targeted group. Also, the researcher took the survey while carrying a small bottle of sanitizer by himself so that he will sanitise the questionnaire before giving him/her questions to answer. The researcher distributed 60 questionnaires, and 57 were returned, resulting in a 95% response rate, which was extremely significant in terms of providing a meaningful sample size from which to draw research conclusions. Saunders and Kalff [20] reiterates that "a research is said to be valuable, reliable and acceptable if at least 60% for the target population provide information for the research".

In order to obtain data from stakeholders such as Harare City Council officials, Environmental Management Agency (EMA) officials, Epworth Local Board (ELB) officials, and Ward Councillors. Interviews were done as planned, with a 100% response rate from interviewees. The researcher employed semi structured interviews whilst conducting telephonic interviews to save time, uphold COVID-19 safety protocols and allows the respondents to respond freely on their time. The researcher first sought for the consent from the local authority in order to carry out his research. Field observations were employed to acquire data that addressed the study's aims and research concerns. Observation aids in grasping the actual data of activities that are taking place on wetlands in this case the activities that are being done by humans that affected wetland provisioning and cultural services. Transect walks were also used to observe the selected sites through a systematic stroll across the community area following a designated path (transect) while listening, and observing the major activities on the wetland. The researcher employed direct observation, which entailed going straight to the field to study human activities on wetlands, with the goal of acquiring first-hand data and allowing the data to be supplemented with images. This guaranteed that the study conclusions were backed up by evidence. As a result, the observer had to interpret what was going on around the Overspill (Ward 6) and try to justify their observations.

Secondary data was gathered from reports, newspapers, textbooks, journals, and other publications for this study. Secondary data was used to find out what are the main human activities that are being done on wetlands in Overspill. The researcher obtained past and current articles that are related to the study. Some of the secondary data was taken from Harare City Council reports, ELB reports and also EMA reports. The researcher analysed old articles that are related to the study in order to find the information that is relevant to his study.

3. Results and discussion

3.1 Socio-demographic characteristics of respondents

The respondents from the distributed questionnaires had a majority of females (65%) whilst 35% were males (**Table 1**). The questionnaire survey results show that the majority of the household respondents (67%) were married a position which also had bearing in terms of wetland use types and benefits derived. In terms of age, the most dominant group was that comprising of 36–45 years old respondents who constituted 35.1%, followed by 26–35 (26.3%), 46 years (22.8%) and above then 18–25 years (15.8%) was the least one (**Table 1**). The dominance (36–45 years) is relevant to the research since it includes of mature persons with job experience who can easily recognise any human activities and affects in wetlands. Research statistics indicate that age respondents of 36–45 years were dominated by economically active people.

Questionnaire survey showed that 29.8% of household respondents attained tertiary education followed by those with Secondary education (O level) (26.3%), then A level (22.8%), and 12.3% were those who never attained any formal education (**Table 1**). The least respondents constituted 8.8% that is primary level. The researcher also collected information on the respondents' educational attainment in order to learn more about their knowledge and awareness of the impact of human activities on wetlands providing and cultural services. Respondents from tertiary level (29.8%) contributed significantly to the validity of the results. The better educational background of the Epworth citizens is crucial as it helps them to understand to have crucial thinking on the impacts of human activities on wetlands provisioning and cultural services.

Questionnaire survey highlights that, self-employed household was the majority of the respondents with 42.1% followed by those who are employed with 33.3% then the none employed attains 24.6% (**Table 1**). Self-employed households were dominant since there is high rate of unemployment in Epworth, hence people tend to engage into different activities influenced water availability in the wetlands. These findings concur with those of Feresu [19] who observed that high rates of unemployment and poverty forced many people to turn to agricultural activities especially on open spaces such as wetlands to improve their incomes and food security. In a study conducted as they indicate that high unemployment forces household to indulge in wetland cultivation for survival.

Questionnaire survey results demonstrated that 33.3% of the respondents reside within a kilometre of the wetlands are the more respondents, followed by those who live 1–2 km (26.3%), and 3–5 km (22.8%) then 5 km and above are the least ones with 17.5%. Respondents who lived a kilometre away from Epworth Ward 6 wetlands played a significant role in the study because they were the ones who are quickly get affected by the impacts of wetland destruction.

3.2 The main land uses on wetlands in Epworth

3.2.1 Agriculture

The results of a questionnaire study revealed that wetlands are used for a variety of purposes. The research findings showed that urban agriculture was the major main land use in Epworth Ward 6 as revealed by 35.1% of the households (**Figure 2**).

	Variable		
1	Gender	Frequency	Percentage
	Male	20	35
	Female	37	65
2	Marital status		
	Single	19	33
	Married	38	67
3	Age		
	18–25	9	15.8
	26–35	15	26.3
	36–45	20	35.1
	46 above	13	22.8
4	Education qualification		
	Primary	5	8.8
	Tertiary	17	29.8
	Secondary (O level)	15	26.3
	A level	13	22.8
	None	7	12.3
5	Employment status		
	None	14	24.6
	Employed	19	33.3
	Self-employed	24	42.1
6	Employment type		
	Temporary	12	21.1
	Permanent	5	8.7
	Full time	14	24.6
	Part time	10	17.5
	Contract	16	28.1
7	Distance		
	1 km	19	33.3
	1–2 km	15	26.3
	3–5 km	13	22.8
	5 above km	10	17.5

Table 1.
Socio-demographic characteristics of the household respondents.

According to the researcher's survey, poverty was the primary motivator for individuals in Epworth Ward 6 to engage in urban agriculture because the wetlands provide rich soil and water for irrigation in their fields. According to ZimStat [16], Ward 6 of Epworth has a higher poverty prevalence rate than the other wards in Epworth.

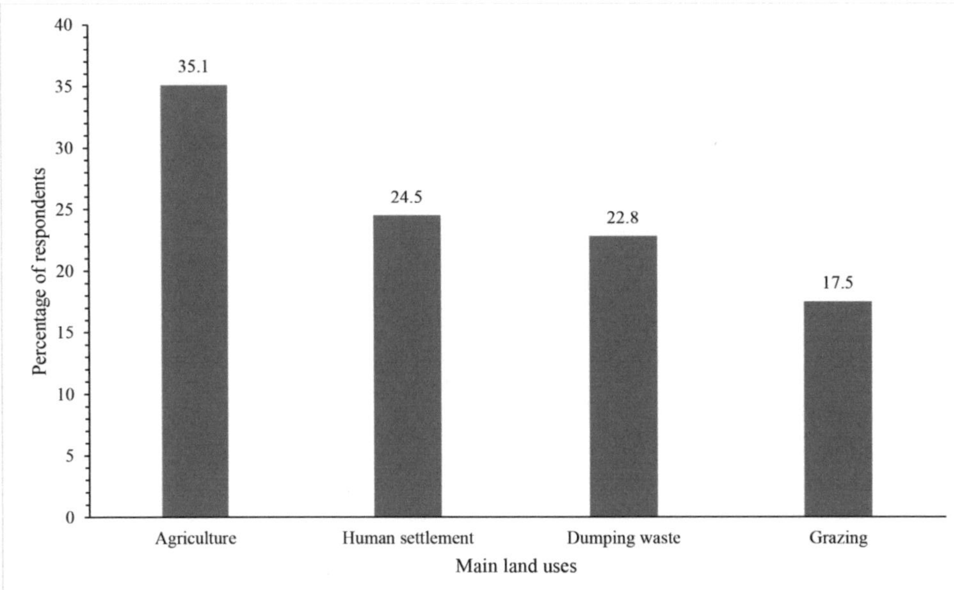

Figure 2.
Main land uses on wetlands in Epworth ward 6.

The researcher also observed that maize, tomatoes and sweet potatoes were the crops that were grown by households in Epworth wetlands. The researcher noted that people in Epworth mostly utilise wetlands in winter season due to the characteristics of seasons, so the surrounding area of wetland is occupied by agricultural activities which end up seeing people encroaching into wetland due to poverty which increase the demand of land for agricultural activities. This demonstrates that wetlands are helping to achieve Sustainable Development Goals (SDGs) 1 and 2, which are to eradicate hunger and poverty, respectively. These findings concur of Feresu [19], who observed that food insecurity, unemployment and poverty has increased the rate of wetland use for agriculture.

3.2.2 Human settlement

Questionnaire survey shows that 24.5% of the households indicated that, apart from urban agriculture, urbanisation was also another major land use in Epworth Ward 6 (**Figure 2**). They noted that Epworth's urbanisation is the outcome of migration, which has resulted in a search for more land to supply housing. One of the Housing Cooperative Heads, in particular, stated that the expanding population as a result of in-migration was putting strain on the land. Housing, according to respondents, can be a significant land use in wetlands. According to the Zimbabwe National Statistical Agency [21], Epworth had a total population of 161,840 people, with 25,597 individuals living in the study area (ward 6). During an interview ELB official highlighted that, shortage of land for accommodation is forcing Epworth Ward 6 residents to build their informal (illegal) settlements stretching into wetlands haphazardly [22] opines that urban growth and rapid urbanisation have been the major drivers diminishing the land resource in sub-Saharan cities.

3.2.3 Dumping of waste

Questionnaire surveys results show that 22.8% of the respondents pointed out that, wetlands were also used for dumping of waste (**Figure 2**). The researcher observed that illegal dumping of garbage and other waste was a threat to the wetland area, which posed several health problems to the community. The researcher also discovered that, the colour of water in wetlands after taking transect walk into field showed that there is a lot of waste that was being dumped in the wetland area. During an interview the EMA officer highlighted that, dumping of waste into wetland is due to the shortage of the resources to the Epworth Local Board to collect garbage on time. As a result, people end up emptying their bins near or open spaces (wetland) thereby affecting wetland quality, its function and also life under water. This conclusion is consistent with Goredema and Sithole [23], who found that people tend to dump their trash in wetland areas because they are considered open and undeveloped.

3.2.4 Grazing

Questionnaire survey highlighted that 17.5% of the households, indicated that grazing was another land use of wetlands in Epworth. Households explained that, the wetland provides palatable pasture and a source of drinking water to livestock including cattle and sheep from the local people. This finding is consistent with that of [16], who state that most rural wetlands, particularly in Zimbabwe, are used for agricultural and cattle grazing, as in Dufuya and Madigane. This is because most temporarily and seasonally waterlogged wetland habitats may provide valuable grazing fields for domestic animals, this decision was made. During an interview the ward 6 Councillor demonstrated that grazing of wetlands was mostly done by farmers who have livestock. These wetlands are used for agricultural purposes as well as animal grazing to keep them healthy. Grazing and agriculture are common uses for wetlands a position also confirmed by Ndlovu et al. [24]. The researcher observed that, people drove their livestock for grazing into wetland areas because that is where they can find good grazing pastures and water for their animals. They did not take any considerations of the importance of vegetation on wetlands because what they care mostly are their livestock. This demonstrates that wetlands are an important element considering the important role that agriculture plays as an important back bone of local communities' livelihoods.

3.3 Factors influencing wetland utilisation patterns in Epworth, Harare

3.3.1 Availability of market for agricultural products

Questionnaire surveys demonstrate that 31.6% of the respondents (**Figure 3**) stated that the availability of market for agricultural products influenced wetland utilisation patterns in Epworth Ward 6 (Overspill). In some circumstances, water removal for agricultural irrigation can amplify the impacts of various pressures on wetland ecosystems, resulting in effects that are greater than those expected from dewatering alone. The households highlighted that, most of the people that utilise wetland for growing of agricultural products are females because a large number of men have migrated in search of formal jobs. Types of crops that are grown in wetlands in Epworth include maize, tomatoes, onions and sweet potatoes, these products were

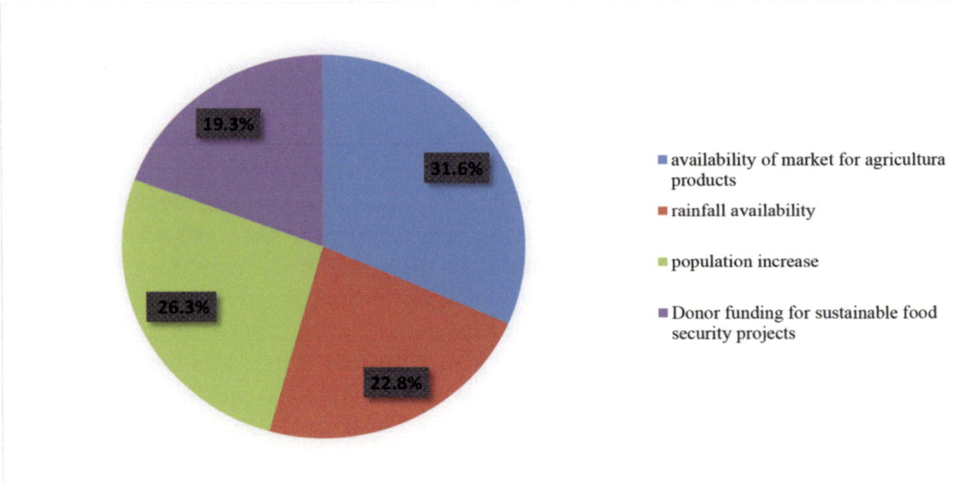

Figure 3.
Factors influencing wetland utilisation patterns in Epworth.

not meant for household consumption only but some are meant for selling since they were at the demand at the daily market to generate more income. Households also explained that, trend in market especially in agricultural sector is pushing most of the people in attaining agricultural activities. As a result, external factors such as market dynamics have an impact on wetland resource demand [24]. So, economic trends have prompted individuals in urban areas to engage in urban agriculture, relying on wetlands for other agricultural inputs such as irrigation water.

3.3.2 Availability of water

Rainfall pattern trends influenced wetland utilisation patterns in Epworth. Research findings highlight that 22.8% of the households pointed out that availability of water is also another major factor that influences wetland utilisation (**Figure 3**). They pointed out that groundwater availability has a significant impact on the nature and function of wetlands. Wetlands are highly dependent on groundwater levels, hence changes in climatic conditions that affect groundwater availability will have a significant impact on the nature and function of specific wetlands, including the types of plant and animal species that live there, analysing groundwater availability as a key driver of wetland changes is an important process [12]. For example, in Matabeleland South located in agro-ecological region 4 with unpredictable rainfall; locals rely on wetlands to absorb rainwater during dry periods [24]. People rely on wetlands to better their livelihoods as a result of poor rainfall. This finding supports Phethi and Gumbo's [25] observation that the availability of water in Makhita has resulted in the expansion of agricultural activities to meet the growing need for food. In these regions, people drain water from these wetlands for their agricultural activities so that they boost their products and income, but this process affect wetland water table through reducing it to a lower level where it will not be able to cope up with the surrounding conditions therefore degraded. Wetlands ecosystem consists of soil, water and vegetation. Changes in hydrologic actions have an immediate impact on the physical condition of a wetland, such as the depth, duration, and frequency of inundation. Human-caused changes in hydrology can have complete control over a wetland's existence and characteristics.

3.3.3 Population increase

Questionnaire survey results show that, 26.3% of the respondents pointed out that, population increase poses serious threat to wetlands (**Figure 3**). During an interview, ELB officials highlighted that in-migration plays a major role in population increase in Epworth which poses serious threats to wetlands natural resources. Population increase has increased the demand of social, economic and environmental services. Human activities such as urbanisation, infrastructure development, recreational, and industrial activities have all posed a threat to urban wetlands [14]. These findings support the findings of Madebwe and Madebwe [26], who claim that population increase, high drought incidence rates, and national and economic development issues led to the establishment of many gardens on the outskirts and within wetlands. One way to relieve pressure on existing land use change is to expand human settlements in wetlands [14].

3.3.4 Donor funding for sustainable food security projects

Questionnaire survey demonstrate that, 19.3% of the household respondents state that, donor funding also plays a role in influencing wetland utilisation (**Figure 3**). Household respondents show that the majority of the donors are Non-Governmental Organisations (NGOs). These donors provide items to local people, which they employ in the implementation of wetland-based food security projects that require the usage of wetlands. Non-governmental organisations (NGOs) provide wetlands fence and distribute farming inputs such as inorganic fertilisers, seedlings, and farm implements. Due to lack of knowledge from the community people end up utilising the donor's resources poorly to an extent that they destroy wetlands. This conclusion is in line with Schuyt's [27] who observed that, wetlands ecosystem is under strain due to poorly implementation of food security projects. This was confirmed by Marambanyika and Beckedahl [5] in a separated study conducted in semi-arid communal areas.

3.4 Provisioning and cultural services obtained by the surrounding community

3.4.1 Provisioning services derived from wetlands by the local community

According to the results of a questionnaire survey, 26.3% of families strongly agree that they get water from the wetlands to support their agricultural activities as well as for domestic usage in their homes (**Figure 4**). They indicated that they grow crop throughout the whole year as long as water is available in wetlands. This finding tallies to that of Dehnhardt et al. [22], who stated that wetlands provide a variety of services that contribute to human livelihoods by providing food and economic opportunities. About 28.1% of the households agree that food is also a major provisioning service that they obtain from the wetland (**Figure 4**). They agreed that the wetlands provide water and food for their cattle as well as the rest of the town. This finding is consistent with that of Svotwa et al. [28], who found that cattle grazing in wetlands has increased in Mwaonazvawo due to the availability of water and pastures on wetlands. According to the findings, 15.8% of households indicate that they also

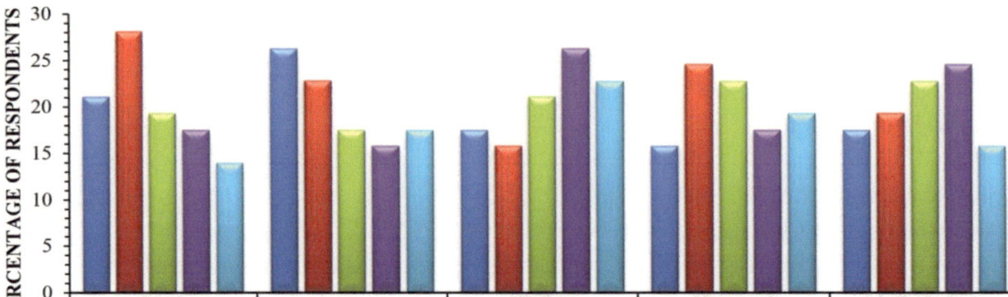

	Food	water	Timber	Firewood	Medical plants
SA	21.1	26.3	17.5	15.8	17.5
A	28.1	22.8	15.8	24.6	19.3
N	19.3	17.5	21.1	22.8	22.8
D	17.5	15.8	26.3	17.5	24.6
SD	14	17.5	22.8	19.3	15.8

PROVISIONING SERVICES

SA A N D SD

Figure 4.
Provisioning services obtained by the surrounding community. Strongly agree—SA, agree—A, neutral—N, disagree—D, strongly disagree—SD.

get wetland materials such as lumber for home roofing and other applications. The respondents indicate that they obtain timber at a minimal scale. Results show that 22.8% of the household's respondents highlight that they neither agreed nor disagreed that they also obtain medical plants from the wetland ecosystem. This is related to the perception that many Epworth Ward 6 households adopt western cultural practices that ignore local tradition. As a result, the majority of the residents in the area rely on clinics or hospitals for medical care.

Results obtained from the survey demonstrate that households' respondents (26.3%) strongly agreed that water is the major resource they obtain from wetlands, because of their proximity to the wetland ecosystem. Most of the respondents lived 1 km (33.3%) from the wetland ecosystem, indicated that most of their activities were done using water for example, irrigation, building of infrastructure and also water for their domestic use since the area is located to a place where there are no water pipes to each and every household. These finding concur with that of Mandishona and Knight [10] who reiterated that wetlands are among the most productive ecosystems, offering households with several opportunities to tap especially for urban inhabitants. Apart from water, the results also show that 28.1% of households agree that they get their food from the wetland habitat. This is because, during their yearly drying, the floodplains become extremely rich in food for aquatic living species, providing a fertile nursery and feeding area for larvae and hatchlings. These findings are also in line with the findings of Madebwe and Madebwe [26] and Mbereko et al. [29], who claimed that during dry periods, wetlands become the only source of moisture, and farmers rely on them for agriculture. Results in **Figure 4** show that households disagree that they obtained timber (26.3%) and medical plants (24.6%).

3.4.2 Cultural services derived from wetlands by the local community

It was established that 19.3% of the households strongly agree that they obtain recreational services whereas 22.8% of the households agree that they utilise wetlands for education and research services (**Figure 5**). Results also show that about 22.8% of the

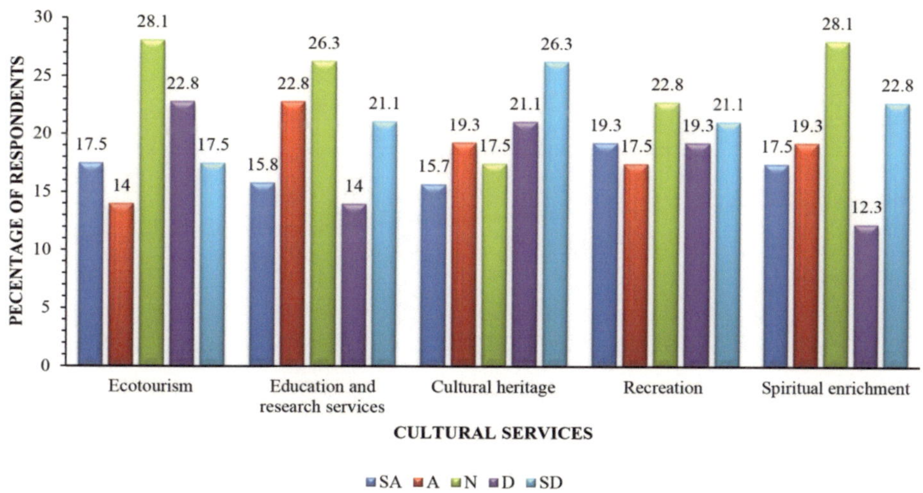

Figure 5.
Cultural services obtained by the community in Epworth. Strongly agree—SA, agree—A, neutral—N, disagree—D, strongly disagree—SD.

household respondents, wetlands give a significant possibility for tourism as well as academic services. Turpie et al. [30] also found that wetlands have educational value since researchers and students from diverse research institutes visit to investigate various elements of wetlands. There was an equal agreement from the respondents as 28.1% neutrally agreed that ecotourism was also a benefit to the community. About 28.1% of the households also highlighted that spiritual enrichment was also another benefit. These findings revealed that the community was already receiving a lot of cultural services (**Figure 5**). During an interview the ward councillor indicated that culture dilution/modification is affecting wetlands in providing numerous cultural services as they were used to be. These concur with the findings of Ongoro [31] who observed that the effects of modernisation, cultural services are rarely observed. As far as conservation was concerned, certain of the beliefs were no longer respected, and some of these cultural values were deemed restrictive. Only 17.5% of the household respondents agreed that they use wetland for spiritual activities such as worshiping through baptism of their worshippers. This finding is consistent with that of [16] who found that wetlands provide spiritual activities on a small scale because such practices were abandoned owing to cultural and technological developments.

3.5 Impacts of human activities on wetland provision and cultural services in Epworth, Harare

3.5.1 Household perception of human impacts on wetlands

The majority of household respondents (31.6%) strongly agreed that agricultural operations have a greater influence on wetlands, according to poll results. Followed by urbanisation (22.8%), then dumping of waste (19.3%) and lastly grazing with 15.8% (**Figure 6**). Household respondents in Epworth indicated that agricultural activities remain the predominant activity over the past decade. Household respondent pointed out that, agricultural activities are the dominant activities that plays a big role in wetland destruction. According to Zimbabwe National Statistics Agency [16], Ward

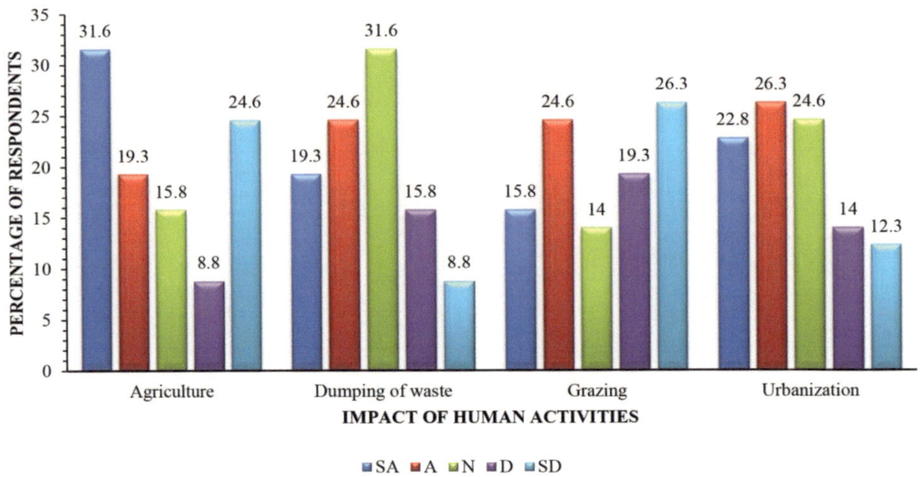

Figure 6.
Major impacts of wetlands in Epworth. Strongly agree—SA, agree—A, neutral—N, disagree—D, strongly disagree—SD.

6 of Epworth has a higher poverty prevalence rate than the other wards in Epworth. During an interview with the ELB official, he stated that since the rate of poverty is high in Epworth ward 6, people tend to engage into agricultural activities. This is because, people wanted to improve their income and household food security. About 19% of the 35.1% of the people aged 36–45 respondent that, due to the availability of market for agricultural products and also availability of water in wetlands pushed many people to engage themselves in agriculture. This is because some of them are among the economically active group that is of 36–45 years. Maize, Onion, Tomatoes and Sweet potatoes are some of the types of crops that are grown into wetlands. Household respondents indicated that these products are being sold to nearer urban areas that are found in Harare (Mbare, Chitungwiza etc). These several agricultural activities in wetlands resulted in their depletion. Household respondents indicated some of the agricultural practices which include, ploughing, tilling, weeding, harvesting and cultivation. This finding concurs with that of [18] who discovered that several agricultural practices which include land clearance through deforestation resulted in wetland degradation.

About 26.3% of the household respondents agreed that urbanisation also have negative impacts on wetlands which affected them in continue providing both provisioning and cultural services. Urbanisation is a process that has increased as a result of population growth, expansion, and the spread of developed structures in a given area. During the interview with Ward 6 Councillor, he reiterates that, population increase has resulted in high demand of land for settlement. So, people end up building their structures into wetland areas therefore affecting the biological and hydrological function of wetland. The poor are the most susceptible in Epworth, according to the researcher, because they lack the financial means to purchase pricey land. As a result, the establishment of informal settlements occurred. This finding is consistent with that of Marongwe et al. [32], who discovered that the impoverished are the ones who build slums and informal communities like Hopely farm and other parts of Epworth.

Questionnaire survey results indicated that, household respondents (31.6%) neither agreed or disagreed that, dumping of waste plays a big role in wetland destruction. Household respondents who live 1 km and 2–3 km from the wetland areas reiterates that, dumping of waste has a role in wetland degradation in Epworth. Other household respondents from areas ranging 5 km away from the wetlands also are among the (31.6%) respondents who neither disagreed that dumping of waste in wetlands has negative impact to wetlands. From the questionnaire results the researcher obtained, household respondents pointed out that, the inconvenience of ELB team that is responsible for carrying garbage's forced many residents to utilise any open space for disposing away garbage's. So, some of the open space include wetland areas. During an interview with the ELB official, he opines that, the ELB has experiencing resource shortages including equipment's in transport so that they can practice their duties in time especially the one of carrying garbage's. From the interview the researcher conducted to EMA, EMA official opines that not only garbage's are thrown into wetland areas, also waste from industries and construction activities are flowing into wetland areas. This finding concurs with that of Goredema and Sithole [23] show that waste is dumped in wetland areas because they are considered open and unutilized.

Questionnaire survey demonstrate that, about 14% of the household respondents neither agreed or disagreed that livestock grazing has a negative result on wetlands. Self-employed household respondents, pointed out that apart from agriculture, urbanisation and dumping of waste, also grazing is continuously contributing to wetland degradation. They opine that, availability of water and green pastures on wetlands drove people (farmers) to come with their livestock's (sheep, cattle and goat) to feed them with grass and watering them. Decrease in wetland vegetation resulting in bare land was attributed to livestock grazing as highlighted by 15.8% of the household respondents who were aged 18–25 years. They demonstrate that, there is an increase in the number of livestock in their ward 6 over the past decade. These all livestock during dry periods, were all fed from wetland products and resulting in bare land. This is because wetlands provide good pastures for livestock. This finding concurs with that of Musamba et al. [33], they recognised that wetlands provide suitable pastures for livestock grazing and are also a constant source of water for livestock, creating a potential condition that could lead to wetland degradation.

3.5.2 Most affected wetland aspect

Questionnaire survey results indicated that about 40% of the household respondents highlighted that water is the most affected wetland aspect in the community followed by vegetation (37%) and soil (23%) (**Figure 7**). Household pointed out that water was affected more because it was used for several activities both social and economic activities. According to Jain et al., [34] the process of runoff from agricultural fields is a major cause of non-point pollution in Indian rivers. As shown on **Figure 3**, availability of water also pushed people in wetland utilisation, this enlighten that water was the major aspect that was being used by the community for agricultural activities mostly. During an interview with the EMA official, he pointed out that water was the most affected aspect since it was being used for a variety of activities. It was being affected by illegal dumping of waste being disposed into it for example sewage (human and animal dump) disposal, pesticides and fertiliser thereby affecting water quality. These findings matched those of Luan and Zhou [35], who claim that pesticide and fertiliser runoff, as well as industrial and municipal wastewater discharges, promote eutrophication in practically all Asian rivers, lakes, streams, and wetlands.

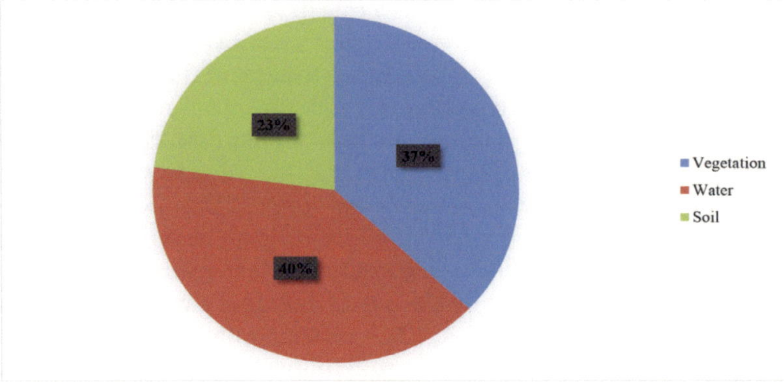

Figure 7.
Most affected wetland aspect.

Questionnaire results demonstrate that, 37% of the household respondents agree that apart from water, also vegetation was also a target aspect from the community (**Figure 7**). Wetland vegetation was being used for several activities in the community as household respondents highlight. Household respondents show that, there are some people in the community who obtain their medical from wetland vegetation, so people utilise vegetation for their medication. During an interview with the ELB official, it was highlighted that, people obtain timber from wetland vegetation which they used for roofing. The Overspill ward councillor revealed that there is an increase in the number of livestock's (cattle, goat and sheep) in his ward since the year of 2000 up to 2022 as his book statistics said therefore which resulted in an increase in bare land from grazing of animals. According to Marambanyika and Beckedahl [5] cattle grazing alters vegetation and causes trampling, which leads to an increase in bare areas. Therefore, wetland vegetation was mostly affected through grazing since there was good pastures for livestock in wetland. This shows that human activities such as cattle grazing have unbearable consequences as they reduced areas of palatable pastures to areas devoid of vegetation.

According to the survey's findings, soil was another factor that affected wetlands for 23% of the household respondents (**Figure 7**). Household respondents highlight that, poor agricultural activities in wetlands led to soil erosion therefore resulting in the decline of soil fertility. During an interview with the elderly household respondents who were above 46 years, shows that anthropogenic activities have caused a dramatic drop in the soil fertility of wetlands over the last decade. Results obtained from an interview conducted by the researcher to ELB officials show that, informal landholders settled in wetland areas used various agents in fertilising the soil. They indicated that, they used these fertilisers with the aim of improving the soil pH. Poor application of these fertilisers led to the loss of soil nutrients hence decline in the soil fertility. From the interview conducted with the EMA official, he respondent by highlighting the impacts of organic and synthetic fertilisers on soil, showing that from the last results they undertake in the field on soil quality, results demonstrate the presence of high nitrogen values. The group concluded that environmental issues resulted from poor nitrogen and phosphorus fertiliser application in terms of soil quality. This finding concurs with that of [5] who believes that various anthropogenic activities in metropolitan areas have resulted in water chemical changes, biodiversity loss, habitat loss, and soil erosion in most wetlands.

3.5.3 Impact of each human activity on wetland provisioning and cultural services

Questionnaire survey results demonstrate that, 40.3% of the household respondents (**Figure 8**) highlight that agriculture has severe impact on wetland which contributes to wetland degradation. The household results show that agriculture has more severe threats to wetlands as compared to other activities as it results in alteration of wetland components such as water and vegetation due to clearance and excessive abstraction. During an interview with the ward 6 Councillor, he stated that agriculture poses more threats to wetland since more people in his ward engage into several agricultural activities. The ward 6 councillor indicated some of the activities that are done in wetlands under agriculture which include, ploughing, planting, weeding, hand crafts, pest and disease control, and harvesting the yields. This finding is similar with that of Feresu [19], he noted that urban agriculture encompasses a variety of agricultural activities, some of which include land clearance by tree removal, resulting in huge deforestation, while irrigation development may also be used to compensate for irregular rainfall patterns. The constant process of agriculture that is increasing year after year as answered, but ELB officials when interviewed said that these cultivation activities exhaust the nutrients in the soil, necessitating fertiliser application to boost yields. This also destroys the soil of wetland from fertile to unfertile therefore reducing the quantity and quality of provisioning services (food, water, etc.) that are being obtained from wetlands. This resulted in food insecurity in the community. So, from the results obtained in the field demonstrated that agriculture was more severe to the destruction of wetlands in Epworth.

Results show that 40.3% of the household respondents revealed that urbanisation has moderate impacts on wetlands, followed by 31.6% (severe), 19.3% (not severe) and the least was 8.8% (no impact) (**Figure 8**). Urbanisation usually negatively impact wetlands both the surface and underground perspectives. Urbanisation include the process of population increase, area expansion and built up of structures in an area. Household respondents (40.3%) revealed that urbanisation has moderate

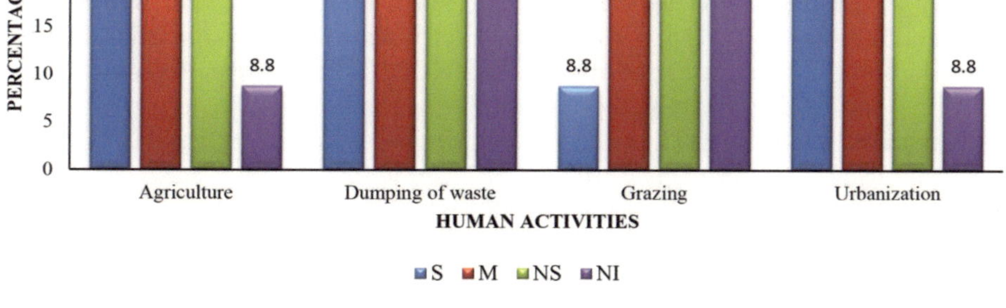

Figure 8.
Impact of human activity on wetland provisioning and cultural services in Epworth.

to wetland soil. Through interviews the researcher conducted to ELB officials and EMA officials, they stated that growing of population in an area or community has resulted into an increase in social demands (land). Several wetland losses have occurred around the world as a direct result of human economic activity [2].

Questionnaire survey demonstrated that 33.3% of the respondent's highlight that, dumping of waste has neither or no severe impact to wetland areas (**Figure 8**). Those who live far away from the wetland pointed out that they did not throw away their garbage into wetland. So, they neither agree nor disagree that dumping of waste into wetland did not have severe impact to wetland since they did not experience their impacts. Goredema [23] opines that resident dispose waste in wetland areas because it has gone some days without being collected. Leaves falling from trees, paper (pampers) and metallic components and also liquid wastes are some of the dump materials that are being disposed into wetland.

The survey's findings reveal that 8.8% of the household respondents pointed out that grazing has posed severe impacts to wetland status (**Figure 8**). During an interview with an EMA official, he pointed out that grazing has severe impacts to wetlands because almost every household that kept livestock, feed their livestock from wetland products during dry seasons. He pointed out that, wetlands face some serious threats from livestock grazing mostly during dry periods because they produce green pastures for livestock. Livestock grazing affect water, vegetation and also soil but at a minimal scale. Livestock grazing leave a bare land which exposes soil to erosion therefore leading to wetland sedimentation. This will affect the quality of food and also quantity of water obtained from wetland. Household respondents indicated that grazing reduces the quantity of medical plants obtained by the community. They also highlight those negative impacts of livestock grazing result in wetland depletion hence affecting recreational activities, ecotourism and spiritual enrichments. Most rural wetlands in Zimbabwe (Dufuya and Madugane) Marambanyika and Beckedahl [5], are used for agriculture and cattle grazing. Several other studies have documented the harmful consequences of cattle grazing on wetlands species, such as the negative effects of trampling on eggs, resulting in habitat loss for wetland species.

4. The nexus between human activities, ecosystem services and the perceived impacts on wetland ecosystems

Results from the present study demonstrate that the majority of the households in Epworth are self-employed a position which pushes then to look for other means to enhance livelihood strategies. Households are therefore left with no option except to turn to wetlands for their survival a position which has also been motivated by the availability of market for agricultural products especially in urban centres such as Harare. Wetland use is also largely driven by other factors such as population increase, rainfall availability and donor funding for sustainable food security projects due to existence of semi-arid conditions. Results also show that the dominant wetland use is agriculture as indicated by most of the households and was also reported to be changing wetland conditions. These specific human activities which are undertaken in wetlands in Epworth were affecting the wetland's ecological components in different several ways which will compromise the ability of the ecosystems to continuously provide goods and services in the near future if they remain unchecked and monitored. These activities were draining water from the wetlands, a situation which would possibly result in wetland degradation and possibly loss which will affect the broad spectrum of goods and services that the wetlands have been providing for instance water for drinking as well as opportunities for eco-tourism and habitat loss. Of all the human activities agriculture and urbanisation severely impacted on the wetland's component such as soil and water. Pertinent to note is the fact that agriculture, urbanisation, dumping of waste and livestock grazing are the main contributors to wetland loss a situation which requires further interrogation through site specific studies. The study advocates for strong support of environmental conservation stakeholders from the government through including these stakeholders on financial budget.it also recommends that EMA and Epworth Local Board to come up with an integrated land use planning approach that will go a long way in addressing issues of land degradation and fragmentation.

The findings show that the wetlands were experiencing habitat loss, biodiversity loss and declining hydrological conditions due to human activities. In light of this it is difficult to maintain wetland conditions if wetland use and access is not regulated and backed by sound environmental practise. Therefore, using the proposed framework in **Figure 9**, there is need to conduct further detailed studies which focus on wetland use activities and their impacts on the wetlands ecological framework using developed frameworks such as the WET-Sustainable Use and WET Health frameworks which inform decisions makers based on a cite of scores and magnitude against the perceived impact in relation to land use activities. This will ultimately contribute to sound decision making as information is based not only with a partial view of reality but is all encompassing.

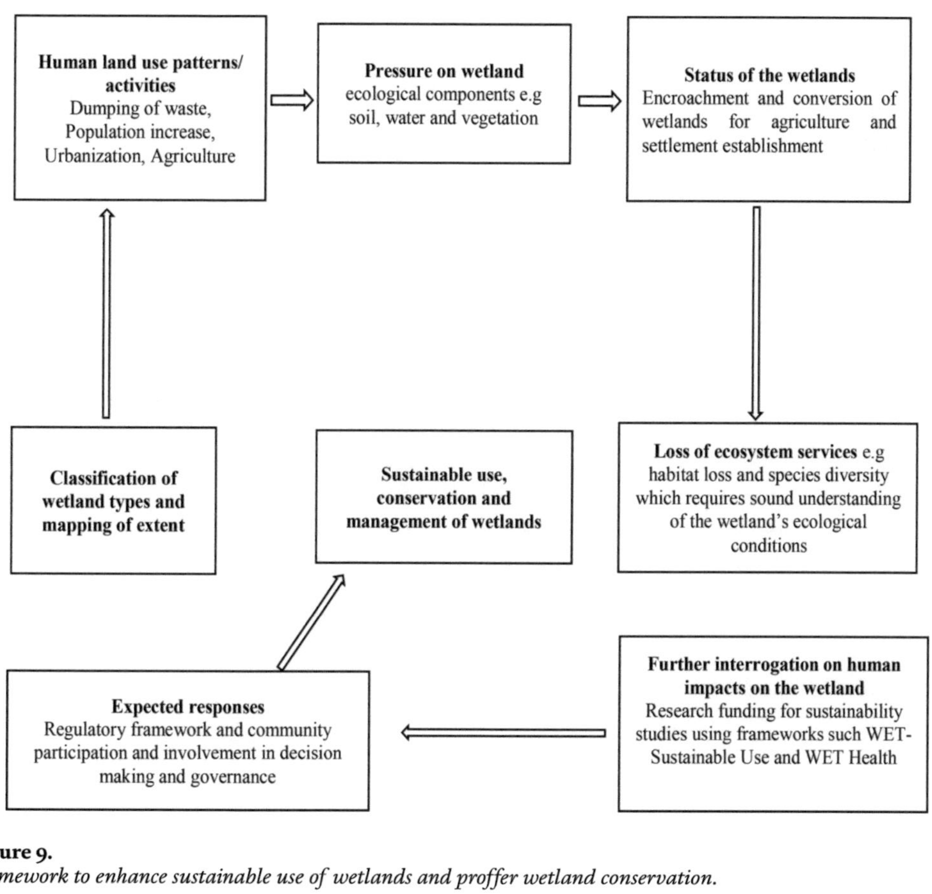

Figure 9.
Framework to enhance sustainable use of wetlands and proffer wetland conservation.

5. Conclusion and recommendations

The article examined the impacts of human activities on wetland provisioning and cultural services in Epworth, Harare. The findings demonstrate that wetlands are mainly used for agriculture activities mainly related to vegetable production followed by human settlement and due to increasing population density being exposed to waste dumping as some households viewed them as waste lands not worthy of management. Households were motivated to utilise wetlands due to the readily available moisture that was mainly used agriculture for agricultural activities for food and income generation. Increased human population was also highlighted as one of the key drivers to wetland use as households were left with no option except to exploit the resources at their disposal. Research findings indicated that local people obtained provisioning and cultural services from the wetlands. It is well known that wetlands provide several services to the surrounding community. Research results demonstrate that ward 6 community obtain several provisioning services from wetlands which include, water, food, timber, firewood. They pointed out that they obtain medical plants from wetlands but at a minimal scale. Household respondents revealed that, there were also some cultural services they obtain from wetland but at a minimal scale. Cultural services obtained by the community include, cultural heritage, spiritual enrichment, recreation, ecotourism and education and research services. These cultural services were now obtained at a minimal scale due to anthropogenic activities taking place in wetlands. This shows that wetlands have been contributing towards the attainment of Sustainable Development Goals 1 and 2 on poverty reduction and zero hunger through sustainable activities within the wetland.

The findings further demonstrate that although wetland have contributed to human livelihoods immensely, human activities such as crop production and livestock grazing have a significant bearing on the provision of good s and services derived. Household indicated that human use activities have a wide range of impacts such as wetland sedimentation, food insecurity, and soil erosion, alteration of hydrological and biological processes of wetland and habitat loss. Households revealed that although they were obtaining some of the provisioning and cultural services from wetlands, the rate is continuously declining, meaning that they are obtaining them at a minimal scale. This was because these wetlands were being used unsustainably hence their depletion. Therefore, in light of this, the Environmental Management Agency and Epworth Local Board should come up with an integrated land use planning approach which is the best way forward in addressing issues that have a hand with human activities and their impact on wetland provisioning and cultural services. This strategy will go a long way towards tackling wetland fragmentation and degradation. Since community participation and stakeholder involvement is key towards the success of any programme, residents surrounding wetlands in Epworth should be educated and encouraged more to engage themselves in practicing Indigenous Knowledge System (IKS) practices for example in steady of using modified fertilisers for improving pH of the soil, it is wise to use inorganic fertilisers such as cow dung or manure etc. Residents surrounding wetland ecosystems should be included in decision making systems which is a good way forward since it reduces conflicts in managing these wetlands.

Author contributions

Conceptualization, Tatenda Musasa (TM); methodology, Kudakwashe Muringaniza (KC), Emmanuel Mhlanga (EM) and (TM). Software, TM; validation, TM and KC.;

formal analysis, TM, KC and EM. investigation, TM, EM and KC, data curation, TM; writing—original draft preparation, TM and KC; writing—review and editing, TM and KC; visualisation, TM, KC, EM. All authors have read and agreed to the published version of the manuscript.

Funding

"The authors declare that no funds, grants, or other support were received during the preparation of this manuscript."

Competing interests

"The authors have no relevant financial or non-financial interests to disclose."

Data availability

"The datasets generated during and/or analysed during the current study are available from the corresponding author on reasonable request."

Author details

Tatenda Musasa*, Kudakwashe C.R. Muringaniza and Emmanuel Dakarai Mhlanga
Department of Geography Environmental Sustainability and Resilience Building, Midlands State University, Gweru, Zimbabwe

*Address all correspondence to: tatendamus14@gmail.com; musasat@staff.msu.ac.zw

References

[1] Springate-Baginski O, Allen D, Darwall WRT. An Integrated Wetland Assessment Toolkit: A Guide to Good Practice. Gland, Switzerland: IUCN and Cambridge, UK: IUCN Species Programme; 2009. xv+144p

[2] Millennium Ecosystem Assessment (MEA). Ecosystems and Human Well-being: Wetlands and Water Synthesis. 2005. pp. 30-38. Available from: http://hdl.handle.net/20.500.11822/8735 [Accessed: March 12, 2019]

[3] Adekola O, Morardet S, De Groot R, Grelot F. The economic and livelihood value of provisioning services of the Ga-Mampa wetland, South Africa. In: 13th IWRA World Water Congress. 2008. p. 24

[4] Chen F, Yao Q. Review of wetland ecosystem services valuation in China. Advance Journal of Food Science and Technology. 2014;**6**(1):1277-1281

[5] Marambanyika T, Beckedahl H. Wetland utilisation patterns in semi-arid communal areas of Zimbabwe between 1985 and 2013 and the associated benefits to livelihoods of the surrounding communities. Transactions of the Royal Society of South Africa. 2016;**71**:175-186

[6] Mitsch WJ, Bernal B, Hernandez ME. Ecosystem services of wetlands. International Journal of Biodiversity Science, Ecosystem Services & Management. 2015;**11**(1):1-4. DOI: 10.1080/21513732.2015.1006250

[7] Ahidur R. Impact of Human Activities on Wetland. A Case Study from Bongaigon District, Assam, India. 2016

[8] Ajibola MO, Adewale BA, Ijasan KC. Effects of urbanization on wetlands. International Journal of Business and Social Science. 2012;**3**(17):310-318

[9] Ramsar Convention Secretariat. Wise Use of Wetlands: A Conceptual Framework for the Wise Use of Wetlands. Ramsar Handbooks for the Wise Use of Wetlands. 3rd ed. Vol. 1. Switzerland: Author; 2007

[10] Mandishona E, Knight J. Users' perceptions and understanding of two urban wetlands in Harare, Zimbabwe. South African Geographical Journal = Suid-Afrikaanse Geografiese Tydskrif. 2019;**101**(3):326-348

[11] Kimani PK. Hydrology and Land use of Lake Nakuru Catchment. The World-Wide Fund for Nature and Egerton University; 1992

[12] Mitsch WJ, Gosselink JG. The value of wetlands, importance of scale and landscape setting. Ecological Economics. 2000;**5**:25-33

[13] Kumar L, Pant KP, Atreya K, Lamsal P. Tourism and Wetland Conservation: Application of Travel Cost and Willingness to Pay an Entry Fee at Ghodaghodi Lake Complex, Nepal. Oxford, UK: Blackwell Publishing Ltd; 2012

[14] Murungweni F. Effect of land use change on quality of urban wetlands: A case of Monavale wetland in Harare. Geoinformatics and Geostatistics: An Overview. 2013;**S1**:1-5

[15] Kitaka N. Phosphorus supply to a shallow tropical lake and its consequences [PHD thesis]. Lake Nairasha, Kenya: University of Leicester, Leicester; 2000

[16] Zimbabwe National Statistical Agency 2012. Zimbabwe Population Census. Zimbabwe Main Report. Harare: Government of Zimbabwe; 2012

[17] Zimbabwe National Statistical Agency 2015. Zimbabwe Poverty Atlas. [Online] Available from: http://www.zimstat.co.zw/ [Accessed: March 20, 2022]

[18] Mutisi L, Nhamo G. Blue in the green economy: Landuse change and wetland shrinkage, belvedere north and Epworth localities, Zimbabwe. Public Administration. 2015;**50**:108-124

[19] Feresu SB. Zimbabwe Environment Outlook: Our Environment, Everybody's Responsibility. Harare, Zimbabwe: Government of the Republic of Zimbabwe. The Ministry of Environment and Natural Resources Management; 2010

[20] Saunders DL, Kalff J. Nitrogen retention in wetlands, lakes and rivers. Hydrobiologia. 2001;**443**:205-212

[21] Zimbabwe National Statistical Agency. Zimbabwe Poverty Atlas. [Online]. 2015. Available from: http://www.zimstat.co.zw/ [Accessed: 20 August 2019]

[22] Dehnhardt A, Häfner K, Blankenbach AM, Meyerhoff J. Valuation of wetlands preservation. In: Oxford Research Encyclopedia of Environmental Science. Oxford University Press; 2019. DOI: 10.1093/acrefore/9780199389414.013.457

[23] Goredema B, Sithole A. Building in Wetlands to Meet the Housing Demand and Urban Growth in Harare. International Journal of Humanities and Social Science. 2013;**3**(8):193-201. [Special Issue—April 2013]

[24] Ndlovu C, Manjeru L. The influence of rituals and taboos on sustainable wetlands management: The case of Matobo District in Matabeleland South Province. International Journal of Scientific and Research Publications. 2014;**4**:1-3

[25] Phethi MD, Gumbo JR. Assessment of impact of land use change on the wetland in Makhitha village, Limpopo province, South Africa. Jamba. 2019;**11**(2). DOI: 10.4102/jamba.v11i2.693

[26] Madebwe V, Madebwe C. An exploratory analysis of the social, economic and environmental impacts on wetlands: The case of Shurugwi District, Midlands Province, Zimbabwe. Journal of Applied Sciences Research. 2005;**1**(2):228-233

[27] Schuyt KD. Economic consequences of wetland degradation for local populations in Africa. Ecological Economics. 2005;**53**(2):177-190

[28] Svotwa E, Manyanhaire IO, Makombe P. Sustainable gardening on wetlands in the communal lands of Zimbabwe. Electronic Journal of Environmental, Agricultural and Food Chemistry. 2008;7(3):2754-2760

[29] Mbereko A, Mukamuri BB, Chimbari MJ. Exclusion and contests over wetlands used for farming in Zimbabwe: A case study of broad-ridge and broad-furrow tillage system on Zungwi Vlei. Journal of Political Ecology. 2015;**21**:322-338

[30] Turpie J, Lannas K, Scovronick N, Louw W. Wetland valuation Vol I-Wetland ecosystem services and their valuation: a review of current understanding and practice. In: WRC Report. Water Research Commission (440/09). 2010

[31] Ongoro W. Assessing the Effectiveness of National Wetland Adaptation and Management Frameworks in Kenya; The Case of Nyando Wetland Ecosystem. Unplished Masters Dissertation, Institute of Environment University of Ottawa. 2017

[32] Marongwe N, Mukoto S, Kuipa T, Chizema R, Chaumba J, Muderere T, et al.

Scoping Study: Governance of Urban Land Markets in Zimbabwe. Harare, Zimbabwe: Urban Landmark; 2011

[33] Musamba EB, Ngaga YM, Boon EK, Giliba RA. Impact of socio-economic activities around Lake Victoria: Land use and land use changes in Musoma Municipality, Tanzania. Journal of Human Ecology. 2011;**35**(3):143-154

[34] Jain CK, Singhal DC, Sharma MK. Estimating nutrient loadings using chemical mass balance approach. Environmental Monitoring and Assessment. 2007;**134**(1-3):385-396

[35] Luan Z, Zhou D. Impacts of intensified agricultural development on marsh wetlands. The Scientific World Journal. 2013;**2013**:1-10. DOI: 10.1155/2013/409439

Chapter 4

Perspective Chapter: How Important is Urban Farming in Indonesia to Support Food Sovereignty?

Arini Putri Hanifa, Eka Triana Yuniarsih, Retna Qomariah, Nurmalinda, Yopi Saleh, Yati Haryati, Indarti Puji Lestari and Susi Lesmayati

Abstract

Urban farming is a simple concept yet significantly impacts food security and food sovereignty for urban households. Indonesian context defined urban farming as cultivation practices, including food crops, vegetables, fruits, herbs, medicinal and ornamental plants, with some combination of fishes and poultry in urban areas, namely home yard, office yard, school garden, communal garden, and many more. This chapter aims to discuss five main topics related to the urban farming movement in Indonesia: (1) The dynamic of yard utilization and food provision policy; (2) The importance of urban farming in society; (3) Community perception and involvement in urban farming; (4) The impact of the pandemic on household food security and food supply chains; (5) Government strategy to sustain participatory urban farming. The sustainability of urban farming still requires government assistance and intervention, and private involvement through corporate social responsibility. The government must support infrastructure both in terms of policy and physical implementation to facilitate the establishment of a network of business partnerships between producer farmers and various market actors in a market chain to step up the era of urban farming industrialization.

Keywords: participatory gardening, home gardening, community garden, urban farming, food security, food sovereignty

1. Introduction

Food security and sustainable agriculture are among the critical issues and priorities in the planning and development in Indonesia. This is in accordance with the global commitments contained in the Sustainable Development Goals (SDGs) to encourage the implementation of sustainable development. One of the 17 the strategy in the SDGs is to end hunger, achieving food security and improved nutrition, and proclaimed sustainable agriculture [1].

The increase in the number of urban population marked by urbanization has become a global issue that requires a comprehensive solution because its impact is related to various aspects of development. Some Frequent challenges and problems that appear among them are not balanced numbers, distribution and population composition, declining environmental quality due to reduced open land/green space, and adequate food availability and quality for city dwellers. One solution that needs to be developed to cope with the problem is urban farming.

Urban farming is a subset of urban agriculture that focuses on income-making agricultural activity. It can be either community farms (driven by social aims) or commercial farms (motivated by profit) and hence can be run as nonprofits or for-profits [2]. Another author characterized urban farming succinctly as food production within cities [3]. In Malaysia, urban farming is described as the practice of planting, processing, and distributing agricultural products in cities and adjacent areas [4]. The same definition is applied in the Indonesian context. Likewise, in this review, urban farming is not seen as an industry but as cultivation practice, including food crops, vegetables, fruits, herbs, and medicinal and ornamental plants in urban areas.

The urban farming movement in Indonesia is promoted to achieve food security and food sovereignty. Food sovereignty is defined as "local people's right to govern their own food systems, including markets, ecological resources, food cultures, and production methodologies." [5]. Food sovereignty is a continuous process in which required adjustments are sought in order to achieve food security [6–8]. It goes beyond the emphasis on food security by giving growers authority [9]. Because of the alleged importance of food sovereignty to self-sufficiency and food security, various nations have enacted legislation to protect this right [10]. Indonesia is one of 10 countries whose constitutions have recognized food sovereignty [11].

2. Dynamics of yard utilization and food provision policy

One of the Indonesian Government's efforts to achieve household food security is to make food available and accessible at the family level by utilizing yard land. Yard utilization offers a solution to anticipate the high rate of rice demand due to population growth through food diversification [12]. The national yard land area is around 10.3 million ha or 14% of the total agricultural land area. If this can be optimally utilized, it will be an enormous potential as one of the providers of nutritious and economically valuable food sources. A yard is a piece of land around the house, whether in front, on the side, or behind the house [13]. The utilization of the home yard is crucial because numerous benefits can be taken. The re-establishment of the yard utilization program in Indonesia with a variety of crop, livestock, and fish commodities is expected to make the food and nutritional needs of household members accessible and affordable.

Through the Ministry of Agriculture, the Government of Indonesia issued various policies related to the importance of food diversification through the optimization of yard land to increase community participation in improving food security and food sovereignty at the household level. Some implemented programs are explained below.

2.1 Sustainable food home area model/model kawasan rumah pangan lestari (MKRPL)

The Sustainable Food Home Area Model (MKRPL) is an implementation of the food security program by the Indonesian Agency for Agricultural Research and Development which was started in 2010. Food security program through optimization of environmentally friendly yard land to meet families' food and nutritional needs. Through MKRPL, public awareness to provide food for families was sown. It is designated as a pilot project to make the home yard productive. Development of MKRPL throughout Indonesia to support food sovereignty and food diversification by involving community participation in rural and urban areas [14].

The concept of m-KRPL/KRPL development is designed in a concentrated area to facilitate management and assistance and provide economic value for the community because it can produce marketable food products. Meanwhile, the sustainable concept is designed with the development of a Village Seedling Garden or communal nursery (**Figure 1**) to supply the needs of KRPL members and communities interested in optimizing yard land and supporting the sustainability of activities [12]. The limited land in urban yards can be addressed by cultivation techniques, namely vertical planting designs, pots, and vines with pergolas—a selection of plants that can be consumed by the family (edible plants). The use of household yards is quite effective with the participation of the community in designing the productive yard [15]. However, some urban residentials in big cities have no space yard for gardening, for this case, a communal/community garden can be a solution. The residents are still able to collaborate in maintaining a garden within shared open spaces (**Figure 2a** and **b**).

The Indonesian Agency for Agricultural Research and Development further developed the KRPL model in 994 regions from 2011 to 2013 throughout Indonesia. The KRPL model by the Central Food Security Agency since 2013 replicates KRPL into P2KP activities in five thousand villages throughout Indonesia. The KRPL program is an effort by the Government through the Ministry of Agriculture to improve food

Figure 1.
A communal nursery in Jakarta.

(a) (b)

Figure 2.
Collaborative work in communal garden (a) seedling preparation (b) maintaining the plants until harvest time.

security and family nutrition. The KRPL program is identical to a similar program called the home garden program developed in many countries worldwide. For example, in Bangladesh, home garden programs can improve food security and reduce malnutrition while reducing poverty in rural areas of Bangladesh [16].

The basic principles of KRPL program development are (i) Utilization of environmentally friendly yards designed for food security and independence; (ii) Diversification of food based on local resources; (iii) Conservation of food genetic resources (plants, livestock, fish); (iv) Maintaining its sustainability through village seed gardens; and (v) Increasing income and community welfare [12].

Although KRPL has been successfully implemented in several regions in Indonesia, however, only a small percentage of Indonesian plan their yards as a food function or grow crops that can meet family food [15]. This condition shows that some people still do not realize the importance of the role of the yard in supporting household's food sovereignty. To optimize the yard utilization, community involvement or people participation needs to be improved, and assistance from relevant agencies for an ongoing basis.

2.2 Sustainable food yard/pekarangan pangan lestari (P2L)

The KRPL program changed its name to the Sustainable Food Yard program in 2020 to expand beneficiaries and use of yard land. The P2L program is conducted to optimize the use of land yard as a source of family food, supporting government programs in handling priority areas for stunting intervention and handling priorities for vulnerable areas of food insecurity, or strengthening food security areas. This activity is conducted through unproductive yards, bare land, vacant land, the land around houses/residential buildings/public facilities, and other environments with clear ownership boundaries, such as dormitories, Islamic boarding schools, flats, houses of worship, and others. The goals and objectives of P2L activities are: (1) To increase the availability, accessibility, and utilization of food for households following the needs of diverse, nutritionally balanced, and safe food (B2SA), (2) To increase household income through the provision of market-oriented food. To achieve these two goals, P2L activities are carried out through an approach to sustainable agriculture development, local wisdom, community engagement, and marketing-oriented [17]. P2L activities help to meet the food needs of families during the current outbreak of COVID-19 [18].

2.3 Sustainable food torch/obor pangan lestari (OPAL)

The Sustainable Food Torch (OPAL) promotes food diversity to fulfill community nutrition through a pilot use of office yard land in 2019. OPAL is a pilot tool for the community in the use of the yard as a source of food and nutrition. This program is conducted within the scope of the Ministry of Agriculture, the Provincial and Regency/City Offices, which organize government affairs in agriculture and/or food to utilize the land around the office area by planting various carbohydrate source commodities, proteins, vitamins, and minerals.

The COVID-19 pandemic has moved the government and the public a lot in increasing the knowledge and skills of urban people about the importance of plant cultivation in urban and suburban areas, to meet the food needs of families during the COVID-19 pandemic. Displaying the model of yard utilization in front of the office complex has attracted visitors and guests to visit and replicate (**Figure 3a**), and become designated outdoor activity and education for young students (**Figure 3b**).

As stated by Handoyo et al. [19], community service activities through the introduction of aquaponics technology can make a positive contribution to urban society. Indicators of success include increased community involvement to participate in vegetable farming with catfish to harvest. Furthermore, the community has benefited from the results of catfish cultivation with an aquaponics system in one cycle amid the COVID-19 pandemic conditions [19]. During the COVID-19 pandemic, the number of locations and community participation in agricultural activities increased, this can be seen from the increase in urban people who grow their vegetables on the vacant lands around the yards [20].

One of the supporting technologies for urban farming during the COVID-19 pandemic that has developed in urban communities is *Budikdamber* (Fish breeding in a bucket with vegetables on the top) as seen in **Figure 4**. Budikdamber often integrates catfish and water spinach, or celery. *Budikdamber* is a development of aquaponics technology, which enables us to cultivate plants and fish on limited lands because this technology does not cut off large areas of land. *Budikdamber* is quite simple to create and maintain, inexpensive, water-saving, and the results can be enjoyed in a relatively short period of time [21]. Aside from *Budikdamber*, fish breeding can be conducted in the ditch in front of the housing complex (**Figure 5**).

(a) (b)

Figure 3.
OPAL at the office yard (a) display of vertiminaponic and wallcaponic (b) Group of young students learn about aquaponic.

Figure 4.
Integration of catfish breeding and water spinach cultivation in a bucket (Budikdamber).

Figure 5.
Tilapia fish breeding in a ditch of housing complex.

3. The importance of urban farming

The importance of urban farming can be viewed from economic, ecological, social, esthetic, educational, and tourism aspects. The benefits of economic aspects include stimulus to strengthen the local economy, increase people's income and

DOI: http://dx.doi.org/10.5772/intechopen.1001601

Figure 6.
Happy faces during harvest time.

Figure 7.
Urban farming in Jakarta.

reduce poverty. If the city community is able to meet their own food needs, more of the city people's money will be used for other interests such as health, education, and housing [22].

When viewed from an ecological aspect, the development of urban agriculture can provide benefits, such as conservation of soil and water resources, improving air quality, creating a healthy microclimate, providing beauty, offers solutions for climate change adaptation.

The social benefits obtained from urban agriculture are increasing food supplies, improving the nutrition of the urban poor, improving public health, reducing unemployment, and reducing social conflicts [22]. Moreover, there is a satisfaction and a pride feeling when we can harvest what we have sown from our yard (**Figure 6**).

The value of health, education, and tourism is obtained from the presence of increased green open space, more CO_2-absorbing areas, improving air quality. Green open spaces become a place to gather, socialize, and recreation and education [23]. For example, in such a big city like Jakarta, the existence of communal garden becomes a place for socializing and spending time for leisure (**Figure 7**).

4. Community perception and involvement in urban farming

The emergence of the COVID-19 pandemic caused many changes in all sectors, and one of them was agriculture. The increase in urban farming enthusiasm in urban communities arises because of working from home and at the same time to survive in large-scale social restrictions implemented in each province and regency/city which narrows the space for movement. With the free time due to work from home and the availability of yard land even though the land is limited, land that was previously less productive can be used as agricultural land to produce household food needs efficiently but healthily [24].

Urban farming is growing along with the increasing interest of people who are influenced by internal conditions requiring activities that can reduce stress and boredom during social restrictions [25]. This urban farming activity is a tool to reduce the negative effects of social isolation due to COVID-19 on a person's emotions and provide positive benefits for food security and dietary quality by increasing fruit and vegetable intake [12]. Most urban farmers feel positive benefits especially psychological benefits [26]. In the environmental aspect, urban farming activities can reduce urban heat islands and air pollution. Regarding nutrition, urban farming activities can improve the food security of households/communities that are food insecure by meeting their intake needs and reducing costs for vegetables or fresh foodstuffs.

During the COVID-19 Pandemic, it has encouraged urban residents to be more active in agricultural activities. Just like the city community in Yogyakarta formed a passionfruit garden during the COVID-19 outbreak in Indonesia. Along with the COVID-19 pandemic, residents became enthusiastic about joining urban farming activities. Agricultural activities in urban areas are a distribution of hobbies and spending free time, where the average urban community are workers and retirees. The COVID-19 pandemic has forced people to stay at home in an effort to stop the widespread spread of the virus, so as to fill the time by cultivating crops, especially vegetables and fruits [27]. The interest of urban people to grow vegetables increased due to the difficulty of the food supply chain, especially from conventional agriculture produced by rural farmers to be marketed in cities. With the use of yard land, community access to food is easier because each sub-district has a vegetable garden that is covered in groups/communities.

The results showed that urban gardening community activity is driven by the desire to reduce household costs, easily get daily food needs, preserve the environment, healthier products, increase income, and fill time free. Urban farming is conducted in various places, such as yards, neighborhoods around houses, and rooftops. To support the principle of reduce, reuse, and recycle (3R) in urban farming, we can make compost in sacks, buckets, drums, used paint buckets, and others. Plants are grown in mineral water bottles, cans, used buckets, pipes, plastic oil packaging, and others [28].

People's interest in urban farming considers aspects of social, environmental, and economic benefits. One example of an area with a high interest in urban farming is Cirebon City. The cultivation of vegetable crops has been carried out with a verticulture system, both verticulture on a vertical and horizontal household scale which contributes to improving household welfare [29].

The social benefit aspect of urban farming was in the highest rank according to the survey results in Pekanbaru City with a score (3.34) and the lowest score is on economic benefits (2.94). Furthermore, respondents' interest in continuing urban farming was at high criteria (51.96%) and remarkably high (13.73). Meanwhile, the perception of the community in Gresik Regency is that urban farming has a positive impact on the environment. Most households (80%) state that urban farming is an appropriate effort in environmental management that pays more attention to sustainable agriculture [30].

Currently, the COVID-19 pandemic has begun to slow down, but the conditions felt by the community because of the COVID-19 pandemic are still being felt and are likely to still be felt for a long time, especially in developing countries. According to Roubík et al. [31] the disruption caused by the COVID-19 pandemic has affected the poor and other marginalized groups, especially those with low purchasing power. The existence of this phenomenon needs to be conducted to mitigate the impact of the pandemic on the entire food system, improve food security, and avoid potential shortages of food. The role of the government in this matter is very necessary to maintain domestic food security after the COVID-19 pandemic. Stimulus is needed in the agricultural sector in the hope of increasing crop production [32]. In addition, strategies that can be implemented to encourage the development of urban horticulture include the provision of technical innovation, organizational innovation, and policy and institutional support [33].

5. The impact of the pandemic on household food security and food supply chains

Food security is a condition of fulfilling sufficient food for the community, both in quantity and quality, at affordable prices. There are three important aspects in food security, as the aspect of food availability or production, the distribution aspect, and the consumption aspect. If food is available but the distribution is disrupted, it will affect community food security, and this happened at the beginning of COVID-19. Social restriction resulted in food distribution being constrained to consumers. Due to the COVID-19 pandemic, the distribution of food was constrained, so the supply of food was reduced, thus threatening people's food security [34, 35]. Many countries have problems in producing food during a pandemic, one of which is due to limited manpower, so the harvest was not optimal [36, 37].

In urban areas, limited land for agricultural development causes most of the food ingredients to be imported from outside the city. Urban areas are highly dependent on food-producing areas, which are located outside cities. However, so far, the urban community's food supply has always been fulfilled, supplied from outside the city and within the city itself. In urban areas, agricultural production is conducted on limited lands owned by the community itself, developers, and the government that have not been utilized.

The social restriction has opened opportunities for companies engaged in online sales, such as: SayurBox and Etanee, SayurBox offers 15 types of vegetables to consumers and Etanee offers 95 types of vegetables and 28 types of fruit [38, 39]. Vegetables that are in high demand in the market are shallots, garlic, potatoes, peeled sweet corn, and several types of leafy vegetables. However, online sales are still limited to urban and buffer zones (Jakarta, Bogor, Depok, Tangerang, Bekasi, and Bandung).

According to the World Bank [40], the COVID-19 pandemic can cause disruptions to the distribution and production of agricultural products, which are also food products for people's needs. COVID-19 caused labor mobility limitations, changes in consumer demand, the closure of food manufacturing facilities, limited food trade regulations, and financial constraints in the food supply chain [41]. The food supply chain was disrupted in countries that imposed a lockdown system to suppress the spread of COVID-19 [42–45]. The lockdown in the city has raised awareness of how crucial it is for residents to have access to food [46]. Movement restrictions, including restrictions on the movement of (agricultural and other) labor and supplies, are likely to disrupt food production as well as food-related logistics and service, posing a challenge to the system's ability to provide enough affordable and nutritious food for everyone [47].

The social restriction has caused many stalls and stores to close and has limited access points for food from agricultural areas' production centers to urban areas. This circumstance causes agricultural production upstream and may cause food prices to fall, immediately affecting farmers, ranchers, and fishers [48]. On the other hand, a lockdown can cause a decrease in food stocks and increase prices due to delays in food distribution [43]. People tend to do stockpiling on food during the lockdown period, increasing the demand for food. In early days of the pandemic, panic buying had caused the run out of food stock in the market.

For example, fulfilling the Special Capital Region of Jakarta Province's food demands is a barometer of national food needs. Because Jakarta is the national capital and the most populous province in Indonesia, with a population of 10,609,700 people (the sixth biggest province in Indonesia) [49], this population directly affects the high demand for food and competition for the use of land resources. Food requirements in Jakarta still depend heavily on food imports from other regions [50].

Jakarta Province engages in inter-regional commerce to address food demands. This business activity creates a distribution network in Jakarta that links food producers with end users and is remarketed to other regions [51]. Food availability depends on the condition of distribution facilities, transportation, and other infrastructure, which will also impact the stability of food prices in Jakarta. Due to transportation issues caused by social restrictions in Jakarta Province, some products, ingredients, or food raw materials are either unavailable or challenging to get. A robust logistical system is required to overcome the long-distance barrier by providing adequate infrastructure (roads, ports) and a modern cold chain system for perishable commodities [48].

Two aspects that can change the food supply chain's long-term effects during and after the pandemic are expanding the online grocery delivery market and customers' preference for "local" food supply chains [52]. In the face of the COVID outbreak, 70% of customers in the United States reduced their food shopping frequency and chose online marketing [53]. People's behavior on internet purchasing has shifted because of COVID-19. During the epidemic, they believed internet purchasing platforms were advantageous and provided convenient access to food and other commodities [54]. Online agricultural commodities marketing platforms may help farmers market their crops. Although the proportion of local farmers using the internet doubled at the start of the 2020 pandemic, only a select few producers and consumers may be able to take advantage of online sales due to the high costs of running online stores or the lack of a reliable internet connection [55].

Due to limited access to imported food, people pick local products as alternatives for necessary food. Local agriculture needs to focus on producing nutritious food like vegetable and fruit products. Moreover, as stated by Schreiber et al. [55], local farmers' adaptability and redundancy helped to maintain the local food chain during the outbreak of a worldwide epidemic. Urban farming is an excellent effort for the movement toward food sovereignty. There is a necessity to promote reflexive consumption by including urban gardeners in urban consumer organizations and to link urban consumers with local producers [56].

6. Government strategy to sustain participatory urban farming

The pandemic gives an important lesson that food availability is the main thing in maintaining food security. During the lockdown, the government encouraged people to stay at home and conduct useful activities at home. Initially, the government encouraged urban farming activities, especially for low-income people where food costs sometimes soared beyond other needs. However, this has changed along with the impact caused by the pandemic. Difficult access to food during the lockdown caused the food system to be disrupted from upstream to downstream and at the same time the availability of local food was not sufficient for the food needs of the local community. The supply of nutritious food is reduced, resulting in fears of food insecurity in the community.

Currently, urban farming is focused on food crops to meet food needs in times of crisis such as a pandemic. Programs related to urban farming activities to the community to foster farmer groups so that they can be adopted and diffused to the surrounding community. The program is designed in dense neighborhoods that do not have large areas of land and then utilized by farming vegetables, fruits, and family medicinal plants, in addition to making ends meet, expected to bring additional income to the family [18].

To encourage community participation in urban farming, the government provides capital assistance in the form of assistance, in the form of vegetable seeds, liquid organic fertilizers, and planting media as well as assistance from local extension workers in the form of technical guidance and motivation. Programs implemented such as *Kawasan Rumah Pangan Lestari* (KRPL), *Lorong* Garden (Longgar), One Village One Variety (OVOV), P2KP, and urban farming activities by NGOs such as Jakarta Gardening, *Kebun Kumara, Kebun Ide*, Sendalu Permaculture, YPBB Bandung, Makassar gardening, and many more groups of people who are concerned about urban farming.

The involvement of the private sector in the form of Corporate Social Responsibility (CSR) is important to sustain the program in society. CSR is defined as the commitment of the business world to contribute to support the economic development process between companies in the local community and the wider community to improve living standards. Development companies that have enjoyed the benefits of the process of transferring paddy fields, and yards into residential areas, and business centers should contribute to the empowerment of local communities. The role of CSR in urban agriculture can be through the concept of developing conservation agriculture, and organic farming in urban agriculture. In addition, CSR can be in the form of financial support for training for farmers, development of superior agricultural land products, pilot plots, and purchase of tools and machinery for agricultural production facilities [57].

For instance, the engagement of PT. Indonesia Power Semarang in urban horticulture by building a greenhouse for farmer groups in Kemijen, Semarang City [58]. Based on the results of the Pertamina Company's CSR program report, replacing hydroponics that initially used Polyvinyl chloride (PVC) pipes into floating rafts with used styrofoam is considered effective and can improve people's welfare in Mariana Village, Palembang [59]. Strategies that can be implemented to encourage the development of urban horticulture include the provision of technical innovation, organizational innovation, and policy and institutional support [33].

7. Conclusion

The importance of urban farming should have encouraged the participation of people in the program. Though the success stories and benefits of urban farming have been admitted, the sustainability of the urban farming movement still requires government assistance and intervention, and private involvement through CSR. The government must support infrastructure both in terms of policy and physical implementation to facilitate the establishment of a network of business partnerships between producer farmers and various market actors in a market chain until these agricultural products reach the consumer level. This is required to step up into commercialization and industrialization of urban farming rather than only self-sufficiency.

Author details

Arini Putri Hanifa[1*], Eka Triana Yuniarsih[2], Retna Qomariah[2], Nurmalinda[2], Yopi Saleh[2], Yati Haryati[3], Indarti Puji Lestari[3] and Susi Lesmayati[4]

1 Research Center for Food Crops, Research Organization for Agriculture and Food, National Research and Innovation Agency, Cibinong, West Java, Indonesia

2 Research Center for Behavioral and Circular Economics, Research Organization for Governance, Economy and Community Welfare, National Research and Innovation Agency, Jakarta, Indonesia

3 Research Center for Horticulture and Plantation, Research Organization for Agriculture and Food, National Research and Innovation Agency, Cibinong, West Java, Indonesia

4 Research Center for Agroindustry, Research Organization for Agriculture and Food, National Research and Innovation Agency, South Tangerang, Banten, Indonesia

*Address all correspondence to: arin010@brin.go.id

References

[1] Handayani W, Nugroho P, Hapsari DO. Kajian potensi pengembangan pertanian perkotaan di kota semarang. Journal of Riptek. 2018;**12**:55-68

[2] Poulsen MN, Spiker ML, Winch PJ. Conceptualizing community buy-in and its application to urban farming. Journal of Agricultural Food System Community Development. 2014;**5**:161-178

[3] Ackerman K. The Potential for Urban Agriculture in New York City: Growing Capacity, Food Security, and Green Infrastructure. New York, USA: Urban Design Laboratory, Earth Institute, Columbia University; 2012

[4] Almalki FA, Soufiene BO, Alsamhi SH, Sakli H. A low-cost platform for environmental smart farming monitoring system based on iot and uavs. Sustainable. 2021;**13**:1-26

[5] Wittman H. Food sovereignty: A new rights framework for food and nature? Environmental Society. 2011;**2**:87-105

[6] Weiler AM, Hergesheimer C, Brisbois B, Wittman H, Yassi A, Spiegel JM. Food sovereignty, food security and health equity: A meta-narrative mapping exercise. Health Policy and Planning. 2015;**30**:1078-1092

[7] Patel RC. Food sovereignty: Power, gender, and the right to food. PLoS Medicine. 2012;**9**:e1001223

[8] Slegers MFW. "If only it would rain": Farmers' perceptions of rainfall and drought in semi-arid Central Tanzania. Journal of Arid Environments. 2008;**72**:2106-2123

[9] Siegner A, Sowerwine J, Acey C. Does urban agriculture improve food security? Examining the Nexus of food access and distribution of urban produced foods in the United States: A systematic review. Sustainability. 2018;**2018**:10

[10] Chihambakwe M, Mafongoya P, Jiri O. Urban and Peri-Urban Agriculture as a pathway to food security: A review mapping the use of food sovereignty. Challenges. 2019;**2019**:10

[11] Knuth L, Vidar M. Constitutional and Legal Protection of the Right to Food around the World. Rome: Food and Agriculture Organization of the United Nations; 2011

[12] Kementerian Pertanian RI. Petunjuk Teknis Optimalisasi Pemanfaatan Lahan Pekarangan Melalui Kawasan Rumah Pangan Lestari (KRPL). Jakarta: Ementerian Pertanian Ri Badan Ketahanan Pangan Pusat Penganekaragaman Konsumsi Dan Keamanan Pangan; 2020

[13] Ashari N, Saptana N, Purwantini TB. Potential of use backyard land for food security. Forum Penelit Agro Ekon. 2016;**30**:13

[14] Aref F. Farmers' participation in agricultural development: The case of Fars province, Iran *Indian*. Journal of Science and Technology. 2011;**4**:155-158

[15] Irwan SNR, Sarwadi A. Lanskap Pekarangan Produktif Di Permukiman Perkotaan Dalam Mewujudkan Lingkungan Binaan Berkelanjutan. Semin Nas Sains dan Teknol. 2015;**1**:1-11

[16] Ferdous Z, Datta A, Anal AK, Anwar M, Khan ASMMR. Development of home garden model for year round production and consumption for improving resource-poor household food security in Bangladesh *NJAS - Wageningen*. Journal of Life Science. 2016;**78**:103-110

[17] Kementerian Pertanian. Petunjuk Teknis Pekarangan Pangan Lestari (P2L). Jakarta: Badan Ketahanan Pangan Kementerian Pertanian; 2021

[18] Rouli Alodia A, Olivia Sitompul A, Kesehatan Masyarakat F. Urban farming selama pandemi covid-19 serta manfaatnya bagi lingkungan dan gizi masyarakat. Journal of Kesehat. 2021;**2**:337-345

[19] Handoyo T, Darsin M, Widuri LI. Kolam gizi akuaponik untuk ketahanan pangan masyarakat urban kelurahan karangrejo kabupaten jember di masa pandemi covid-19. Panrita Abdi-Jurnal Pengabdi. 2022;**6**:114-122

[20] Murdad R, Muhiddin M, Osman WH, Tajidin NE, Haida Z, Awang A, et al. Ensuring urban food security in Malaysia during the COVID-19 pandemic — Is urban farming the answer? A review. In: Sustainability. Switzerland: MDPI; 2022;**14**(7). DOI: 10.3390/su14074155

[21] Noviana F, Fadli ZA, Hastuti N. Sosialisasi budikdamber sebagai salah satu upaya penguatan ketahanan pangan rumah tangga selama masa ppkm akibat pandemi covid-19. Harmon. J. Pengabdi. Kpd. Masy;**5**:68-73

[22] Setiawan B, Rahmi DH. Ketahanan Pangan, Lapangan Kerja, dan Keberlaniutan. Kota: Studi Pertanian Kota di Enam Kota Indonesia; 2004

[23] Fauzi AR, Ichniarsyah AN, Agustin H. Pertanian perkotaan: urgensi, peranan, dan praktik terbaik. Journal of Agroteknologi. 2016;**10**:49-62

[24] Karyani T, Djuwendah E, Sukayat Y. Pemberdayaan Masyarakat Di Masa Pandemi Melalui Pertanian Organik Di Lahan Pekarangan Kawasan Perkotaan Jawa Barat. Dharmakarya. 2021;**10**:139

[25] Bulgari R. The impact of covid-19 on horticulture: Critical issues and opportunities derived from an unexpected occurrence. Horticulturae. 2021;**2021**:7

[26] Salman TM, Matloobi M, Wang Z, Rahimi A. Understanding the dynamics of urban horticulture by socially-oriented practices and populace perception: Seeking future outlook through a comprehensive review. Land Use Policy. 2022;**122**:106398

[27] Suryantini A. Perceived Benefits and Constraints in Urban Farming Practice during COVID-19. IOP Conference Series Earth Environmental Science. 2021;**2021**:686

[28] Wicaksono FY, Nurmala T. Respons Peserta Pelatihan Urban Farming Yang Ramah Lingkungan Secara Online. Dharmakarya. 2021;**10**:154

[29] Wachdijono WS, Trisnaningsih U. Penerapan Urban Farming " Vertikultur " untuk Menambah Pendapatan Rumah Tangga di Kelurahan Kalijaga Kecamatan Harjamukti Kota Cirebon *Pros*. Seminar Nasional Unimus. 2019;**2**:374-381

[30] Purba S, Amir IT. Household perception of urban farming for environmental coastal area HouseholdB (Case Study District Gresik). IOP Conference Series Earth Environmental Science. 2021;**2021**:695

[31] Roubík H, Lošťák M, Ketuama CT, Procházka P, Soukupová J, Hakl J, et al. Current coronavirus crisis and past pandemics - what can happen in post-COVID-19 agriculture? Sustainable Production Consumers. 2022;**30**:752-760

[32] Churiyah M, Dharma BA. Optimalisasi Pemberdayaan Kelompok Urban farming untuk meningkatkan ketahanan pangan pasca pandemi covid

19. Portal Riset dan Inovasi Pengabdian Masyarakat (Prima); 2022;**2**(1):79-89

[33] Sastro Y. Pertanian perkotaan: peluang, tantangan, dan strategi pengembangan. Bul. Pertan. Perkota. 2013;**3**:29-36

[34] Mouloudj K, Bouarar AC, Fechit H. The impact of Covid 19 pandemic on food security. Les Cah du Cread. 2020;**36**:159-184

[35] Siche R. What is the impact of COVID-19 disease on agriculture? Science Agropecu. 2020;**11**:3-9

[36] Gregorio G, Ancog R. Assessing the impact of the COVID-19 pandemic on agricultural production in Southeast Asia: Toward transformative change in agricultural food systems. Asian Journal of Agricultural Development. 2020;**17**:3-51

[37] Dev D, Kabir K. COVID-19 and food security in Bangladesh: A chance to look Back at what is done and what can Be done. Journal of Agricultural Food System and Community Development. 2020:1-3

[38] Sayurbox. Big Alpha - Mengenal Sayurbox. Alternatif Belanja: Sayur dan Buah Online; 2023

[39] Venny S. Penjual sayuran online alami permintaan berlipat-lipat saat pandemic. Jakarta: Kontan; 25 Apr 2020

[40] World Bank. A Shock like No Other: The Impact of COVID-19 on Commodity Markets. Washington DC: World Bank; 2020

[41] Aday S, Aday MS. Impact of COVID-19 on the food supply chain. Food Quality Safish. 2020;**4**:167-180

[42] Abhishek BV, Gupta P, Kaushik M, Kishore A, Kumar R, Sharma A, et al. India's food system in the time of covid-19. Economical Political Weekly. 2020;**55**:12-14

[43] Fawzi NI, Qurani IZ, Rahmasary AN. COVID-19: Implication to Food Security. Jakarta. 2020. DOI: 10.13140/RG.2.2.15160.90888

[44] Hobbs JE. The Covid-19 pandemic and meat supply chains. Meat Science. 2021;**181**:108459

[45] Chari F, Muzinda O, Novukela C, Ngcamu BS. Pandemic outbreaks and food supply chains in developing countries: A case of COVID-19 in Zimbabwe. Cogent Bus Man. 2022;**9**:2026188

[46] Pulighe G, Lupia F. Food first: COVID-19 outbreak and cities lockdown a booster for a wider vision on urban agriculture. Sustainable. 2020;**12**:5012

[47] Bakalis S, Valdramidis VP, Argyropoulos D, Ahrne L, Chen J, Cullen PJ, et al. Perspectives from CO+RE: How COVID-19 changed our food systems and food security paradigms. Current Research Food Science. 2020;**3**:166-172

[48] Putri NA, Sitompul FR. Smart food supply chain: Recommendations after COVID-19 pandemic in agricultural industry. In: Ttialih R, Wardiani FE, Anggriawan R, Putra CD, Said A, editors. Indonesia Post-Pandemic Outlook: Environment and Technology Role for Indonesia Development. Jakarta-Indonesia: Overseas Indonesian Student's Alliance & BRIN Publishing; 2022. pp. 209-225

[49] BPS. STATISTIK Indonesia 2022. Jakrta: BPS Indonesia; 2022

[50] Anugrah IS, Saputra YH, Sayaka B. Dampak Pandemi Covid-19 Pada Dinamika Rantai Pasok Pangan Pokok Dampak Pandemi Covid-19:

Perspektif Adaptasi dan Resiliensi Sosial Ekonomi Pertanian. 2020;**3**:297-319

[51] Sulaiman AA, Jamal E, Syahyuti KIK, Wulandari S, Torang S, Hoerudin BF, et al. Menyangga Pangan Jakarta: Sebuah Konsep Keterkaitan Pangan Kota Besar dan Wilayah Peyangga. Jakarta: IAARD Press; 2017

[52] Hobbs JE. Food supply chains during the COVID-19 pandemic. Canadian Journal of Agricultural Economics. 2020;**68**:171-176

[53] Barman A, Das R, De PK. Impact of COVID-19 in food supply chain: Disruptions and recovery strategy. Current Research Behavioural Science. 2021;**2021**:2

[54] Shahzad MA, Razzaq A, Qing P, Rizwan M, Faisal M. Food availability and shopping channels during the disasters: Has the COVID-19 pandemic changed peoples' online food purchasing behavior? International Journal of Disaster Risk Reduction. 2022;**83**:103443

[55] Schreiber K, Soubry B, Dove-McFalls C, MacDonald GK. Diverse adaptation strategies helped local food producers cope with initial challenges of the Covid-19 pandemic: Lessons from Québec, Canada. Journal of Rural Studies. 2022;**90**:124-133

[56] Nasution Z. Indonesian Urban Farming Communities and Food Sovereignty. 2015

[57] Muljawan RE. Kontribusi Perusahaan Pengembang Terhadap Pemberdayaan Petani Perkotaan dalam Mewujudkan Ketahanan Pangan. Pros. Konf. Nas. Pengabdi. Kpd. Masy. dan Corp. Soc. Responsib. 2019;**2**:1478-1483

[58] Wulandari R, Martuti NKT, Syamsul MA, Mutiatari DP. Peran Program CSR PT. Indonesia Power Semarang PGU Dalam Mendukung Pertanian Perkotaan di Kelurahan Kemijen. Journal of Abdimas. 2021;**25**:225-232

[59] Hilmy SI, Poetri IDM, Putri BA. Efektifitas Program CSR Hidroponik Rakit Apung Terhadap Peningkatan Taraf Hidup Penerima Manfaat Program. Ampera: A Research Journal of Political Islam Civilization. 2022;**3**:156-164

Chapter 5

Therapeutic Impact of Engagement in Green Spaces

Kasey Stepansky, Theresa Delbert and Janet C. Bucey

Abstract

Active engagement in green spaces has been shown to improve physical, mental, and social well-being. Blending the topics of forest bathing, therapeutic sensory gardens, and nature meditations, this chapter will unpack the therapeutic effects of active and passive engagement in green spaces. Frequent exposure to and engagement in green spaces has been found to decrease feelings of anxiety, social isolation, and stress levels. Spending time in green spaces can promote restoration and offer recovery from daily stressors. With the perspectives of the authors' expertise in holistic health as occupational therapy practitioners, the chapter will present how the influence of usage and dosage of green spaces affect stress, social isolation, and a sense of well-being to facilitate occupational balance (flow). This chapter will report how green spaces have been assessed in the current evidence and the questions that remain regarding the impact on personal, group, and population well-being.

Keywords: green space engagement, therapeutic impact, dosage, occupational balance, well-being

1. Introduction

A growing number of researchers [1–3], mental health advocates [4, 5], and media outlets [6, 7] support engagement within nature and green spaces to promote well-being. The purpose of this chapter is to explore the evidence on the therapeutic impact of engagement in green spaces and supplement the review of the evidence with the authors' experiences piloting a therapeutic sensory garden intervention to support the well-being of university students.

As occupational therapists, the authors consider well-being within the context of a person's ability to engage in activities that are meaningful to them. Participating in activities that are meaningful can enhance a person's physical, mental, and social life. Occupational therapy training is holistic in bridging science and art by adapting, creating, and modifying activities to optimally support the person within their environment [8]. From the inception of occupational therapy as a profession in 1917, its founders viewed people as inextricably tied to their environment [9]. From this holistic lens of a person within their environment, the profession connects deeply with the impact of green space on individuals' well-being. Records of occupational therapists using gardening as a therapeutic intervention are recorded as far back as 1932. The natural environment has been a context for meaningful engagement in activity, production, and socialization [10].

A common concept within occupational therapy is **occupational balance**. The occupational therapy practice framework defines occupational balance as "the proportion of time spent in productive, restorative, and leisure occupations" (p. 12) [8]. Occupational balance is an outcome of health and wellness that ensures that the demands of different activities meet the needs of the individual [8, 11]. In 2021, the definition of occupational balance has expanded outside of a Western paradigm utilizing influences from Chinese culture to address humanity's need for harmony [12]. This philosophy on occupational balance follows the traditional Chinese belief that each human ultimately has the goal of being in unity with nature [13]. Eriksson et al. [14] investigated the impact of using therapeutic gardening to promote occupational balance. The described themes experienced by participants included being absorbed in the present and accepting self as connections to the protective environment of the therapeutic garden [14]. These findings intersect with the basis for this chapter by describing what, how, and why of green space as a means for therapeutic impact.

This chapter is meant to further the discussion of the therapeutic utilization of green space by using references within and outside of the occupational therapy discipline. Intentional questions will be threaded throughout the chapter related to what makes green space therapeutic.

- What defines green space? What has been studied and how does it differ depending on locality, type, and seasonal use?

- How can green space be used therapeutically? What are the intentional means of intervening within green space?

- Why do we use green space? What are the outcomes and impacts of the green space on well-being?

2. Typology of green spaces that impact well-being

Green space has been defined as public ground covered primarily by grass, trees, plants and other vegetation that is open and accessible. The openness of this land allows for recreation and provides places to view and experience nature [15]. Natural green spaces have been connected to small urban parks and larger wilderness areas such as protected national parks or nature preserves [1]. Additional descriptions have included gardens, outdoors, eco-roofs, green infrastructure, wilderness, greenery, and forests [2, 3, 16]. There is no universally agreed definition of green space, and in order to collect evidence, more specific delineation of spaces can assist with comparing impacts of green space on wellbeing [2]. Providing greater detailed descriptions with photographs, maps, and written overviews to the type of green space can provide more comparable reviews of impact of green space on well-being [2].

The design of green space for therapeutic purposes needs to have components for pathogenesis and salutogenesis. Pathogenesis design focuses on prevention or management of disease, while salutogenesis looks to create and enhance optimal health [17]. Considerations for both purposes are vital to the design and list environmental qualities to include considerations of access, orientation to the space, richness in species, serenity for reflection, and opportunities for social connection [17]. The natural environment is restorative when it is immersive and engaging in order to remove the individual from their typical context and everyday stressors to allow for an experience of escape [3]. Further, with Kaplan and Kaplan's "Attention Restoration Theory", green spaces are intrinsically therapeutic due to the decreased cognitive and attentional demand compared to other environments [3, 17, 18].

Green spaces have been defined by different types/green infrastructures with similar benefits to well-being. The definitions and delineations of these spaces have overlapping characteristics which can further conversations about what is integral to the space to maximize therapeutic benefit.

2.1 Garden green spaces

Gardens are a common green space utilized for active engagement in nature. Types of gardens could be individual domestic gardens, community gardens, healing gardens, and therapeutic sensory gardens. Defining characteristics of a garden include planned space for physical movements, cultivation, enjoyment of display, and closeness to nature [19]. Gardens are planned spaces that should address the publics' accessibility, aesthetic value, maintenance, and safety through appropriate design [19]. Therapeutic gardens, as defined by the American Horticulture Therapy Association, are "A garden designed for use as a component of a treatment, rehabilitation, or vocational program" (p. 3) [20]. Expanding on therapeutic gardens, therapeutic sensory gardens provide an environment tailored to engage all of the individuals' senses including sight, taste, smell, touch, hearing, and body movement [21].

Gardens as a therapeutic green space have been used in China since the year 717 AD when the first public hospital was created with green settings located in the form of courtyards. Chinese monks would care for individuals with illnesses and disabilities in more remote rural settings where the natural surroundings were considered part of the palliative care [22]. Western counties have histories of use of healing gardens as part of hospital settings since the 1800s, however in the early 20th century during the industrial revolution many green spaces were removed from health care facilities due to increases in high-rise building designs. The revival of healing gardens within healthcare settings in the West did not return until the 1990s [22].

2.2 Urban mixed green spaces

Within urban spaces, the types and settings of green spaces connected to supporting mental health include urban forests, neighborhood green space, parks, and eco roofs [23, 24]. Urban planners often consider multiple factors when designing the space including environmental factors (improving air quality, absorbing noise/creating soothing soundscapes, visual stimulation, mitigating heat), space for activity (physical and social), and intentional design for social cohesion (fostering social capital, neighborhood satisfaction, sense of security, belongingness to space) [23, 25]. Indicators have found based on mental health needs that communities prefer green spaces with moderate vegetation complexity and a tree canopy over a simple open space of grass [23]. Urban green spaces must also consider potential environmental determinants to the use of the space including accessibility, quality and condition of features, and feelings of safety within the environment [26].

Green Space Design of a University Therapeutic Sensory Garden

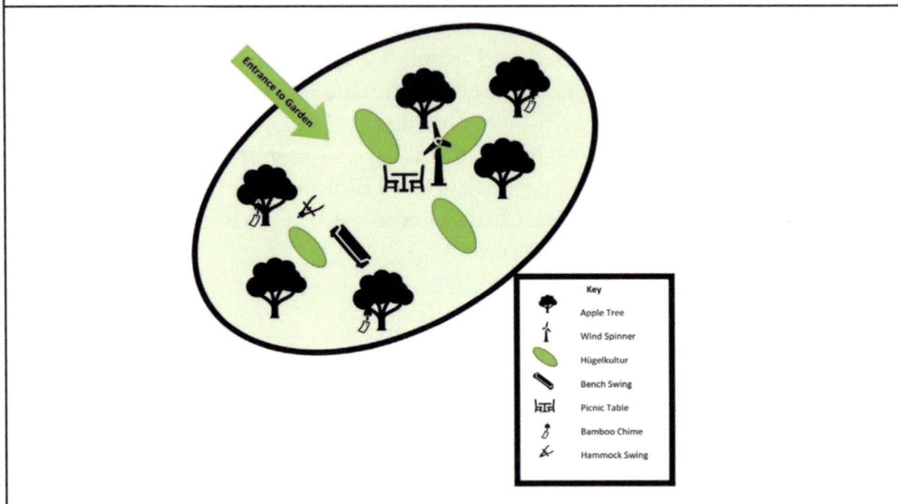

The above design graphic illustrates the design planning of a University Therapeutic Sensory Garden. The authors had access to build the garden space within a 388-acre LEED Platinum certified, sustainable campus that provides elements of both green spaces described as gardens and rural wilderness. The campus is within rolling woodlands, fields, and closely connected to an organic farm. With the components of design known to be connected to well-being including serenity, engagement in individuals' senses, removal from outside distractions (sense of escape), sustainable lushness of vegetation choice, the design of the space was a collective effort between the researchers, invested stakeholders, and the campus community [30].

Figure 1.
Case study: green space design [30].

2.3 Rural wilderness green space

The key features of rural wilderness spaces associated with improvements in well-being are less structured and defined as compared to garden and urban green spaces where planning leads to more descriptive outlines/graphics and examples. Studies looking at the impact of wilderness on wellbeing have labeled the space as broadly as unfamiliar outdoor wilderness settings [27] to increased specificity related to classes of the natural land cover including: woodland, farmland, grassland, mountain, bog, coastal with either freshwater or saltwater [28]. Skärbäck et al. further defined rural environmental qualities for well-being to include the following characteristics: serene (limited sound, natural view), wild (plants self sown/old paths), lush (richness in species), spacious (removed, enabling feelings of escape) (**Figure 1**) [29].

3. Therapeutic interventions in green space

Green space and nature-related interventions are utilized to improve a sense of well-being across therapeutic and healthcare disciplines. What is included in the nature experience/activity varies depending on discipline and active ingredients of the intervention [1, 3]. Context, population, and culture can influence how and when nature based interventions are used [3]. Four types of green space therapeutic interventions are addressed in this section: Horticultural Therapy, sensory gardens, nature meditations, and forest bathing.

3.1 Horticulture therapy

Of particular historic interest and predominance in nature-based interventions is Horticulture Therapy. Horticultural Therapy is a defined discipline supported by associations in countries all over the world including Canada, Taiwan, Australia, Peru and the UK [31]. As a therapy, it is known to have roots in psychiatry and occupational therapy particularly following WWII, as patients demonstrated significant improvement in mental health utilizing task engagement with plants and nature to address designated treatment goals [32]. The occupation of horticulture was considered therapeutic in the early 20th century primarily due to the ability of medical patients to be outside of the sterility of a medical institution, and also a means for raising funds by selling produce and flowers [33]. Horticulture therapy developed into an independent discipline in the latter half of the 20th century [32]. Horticulture could be a beloved past time steeped in familial ties, cultural relevance, and cherished memories or depending on the person could bring fear and apprehension due to lack of knowledge and experience [33]. Meaning and emotional attachment placed on gardening could vary however common themes for populations that have experienced benefits to their well-being through this therapeutic medium include:

1. The beauty of the natural environment intrigues, relaxes, and provides an opportunity for reflection on life concerns.

2. The cultivation of nature is intrinsic to supporting the natural world and the planet's life.

3. Nurturing and attending to the growth of plants can create a feeling of relationship and closeness.

4. Sharing experiences with others in the gardening space supports community building [1, 3, 10, 21].

3.2 Mindfulness in nature

Green space meditation and mindfulness interventions demonstrate positive changes in physical, psychological and pro-social behaviors [34, 35] and are regularly used in treatment of mental health challenges and illnesses. More recent connections in literature report that when mindfulness meditation practices utilize green spaces or nature support, that the positive impact is greater for addressing mental health issues such as depression, rumination, and stress [34]. These meditation practices in green space can be described as introspective awareness or external focus on the present moment. The primary effort of these practices are to connect with the present moment, and reduce the cognitive state of a wandering mind or mindlessness [35]. The cognitive engagement strategies utilize nature as the foreground and an active ingredient for mindful practices [36]. Connectedness to nature is an integral component of meditation practices. As an individual increases the prominence of nature memories the impact of connectedness is lengthened for an individual's sense of well-being [34–36]. The individual in the meditative state feels a part of a larger universal state of being as a partner too and member of the natural world [36]. The connection of mindfulness within the green spaces have also been linked to person, group, and population valuation of pro-environmentalism [36].

3.3 Forest bathing

Originating in theory from Eastern cultures, forest bathing (Shinrin-yoku) is walking and breathing in nature to elicit positive psychological and physiological benefits [37]. Forest bathing involves individuals walking slowly within forested nature spaces with a focus on breathing methods and heightened awareness of the natural environment around the individual [37]. The Forest Agency of Japan initiated this wellness initiative in 1982, identifying forest bathing to combat stress and reduced attention caused by fatigue. Since then Japan has been the leader in researching and defining forest bathing as effective for youth and adults in reducing anger, fatigue, depression, confusion and improving positive emotion [37–39]. Further investigations such as a study in Eastern Europe, suggest that winter forest bathing when the leaves are off the trees is also effective in positive emotions, restoration and vitalization [40]. Some reports have been able to identify the significant difference in urban walking to forest walking, giving support to the need for forest bathing for the increasingly urbanized population and for use in public health [39].

"Active ingredients" in the University Therapeutic Sensory Garden

Therapeutic Sensory Garden Elements.

Senses	Sensory elements included in the Therapeutic Sensory Garden
Sight: *Visual*	This included an array of colors throughout the garden by incorporating the following: • Trees, plants, and hügelkultur garden beds (apple trees, salvia, lemongrass, mums, butterfly bushes, gaura, goldenrod, Helen's flower, blue iobelia, tickseed sunflower, and hummingbird mint) • Solar powered lights • Wind spinner
Sound: *Auditory*	This included sounds of: • Water trickling from fountain • Birds chirping • Insects flying around
Touch: *Tactile*	• Lamb's ear and sage • Water within the fountain • Course texture of the mulch • Rough rigid feeling of the tickseed sunflower
Taste: *Gustatory*	• Apples • Mint Leaves • Lavender • Thyme
Smell: *Olfactory*	• Lavender • Hummingbird Mint • Thyme
Movement: *Vestibular*	• Swing • Open area to walk, run, dance, climb trees, etc.

Engaging in sensory experiences offered in the outdoor garden that is located in the orchard of a 388-acre university farm resulted in positive emotions, connectedness and sensory responsiveness. The participants in this case study visited the therapeutic sensory garden one hour per week for 5 weeks [30].

Figure 2.
Case study: active ingredients [30].

Overall, however, the length, duration, and types of activities, in use in forest bathing are still being determined as well as the most impacted populations for this nature-based intervention [37–40].

3.4 Sensory modulation in a nature space

Green spaces inherently provide sensory experiences including visual stimuli, variety of touch, movement, sounds and smells. Walking on the ground provides gravitational pull that is countered by muscles and joint sensory adaptations. A person's ability to take in sensory stimuli and process it optimally for adaptation with sensory integration theory is termed sensory modulation [41]. This relies on an active engagement of the person within the environment of stimuli and assumes that an individual will seek out a variety of stimuli for the duration needed in order to achieve a sense of well-being from successful sensory modulation [30]. Options to engage in sensory modulation are intrinsically designed within therapeutic sensory gardens with populations having the ability to modulate their sense of emotional and physical well-being by choosing to engage in different sensory experiences that are available and easily accessed sensory choices [19, 42]. **Figure 2** highlights active ingredients of the designed sensory garden created by the authors.

4. Therapeutic outcomes: Impact of engagement

Green space engagement provides a variety of outcomes for individuals, groups, and populations. Evidence demonstrating the effects of nature on psychological or physiological health and well-being includes longitudinal studies, experimental studies, observational studies, controlled laboratory studies, and studies done in nature itself. Outcomes from time spent in nature have come from self-reports, psychological measures or assessments, and the assessment of biomarkers or measurable physical substances (such as blood pressure or temperature).

4.1 Passive/active engagement

Nature can serve as the context, or the background in passive engagement. In this way, passive engagement refers to being in, or simply experiencing nature [43]. This can include listening to birds, smelling flowers, feeling sunshine, or looking at a landscape. Positive health outcomes can be felt from passive engagement [44–48]. Of particular interest, passive engagement does not always have to be in nature for positive effects to be felt. For instance, passively viewing or hearing images and sounds of nature can have a positive impact on one's psychological and physiological health [48]. Viewing nature from a window can result in fewer workplace ailments and higher job satisfaction overall [47].

Nature can also be a partner, a collaborative means or ends. This type of active engagement refers to doing in nature [43]. Active engagement may involve participation in activities such as exercising, socializing, gardening (including weeding, sowing, harvesting, picking flowers, etc.), or tending to animal life such as birds. It is a holistic process of both mind and body with intentional use of natural space [30].

Environmental quality	Description	Active	Passive
Joyful and meaningful activities	Engagement in enjoyable, preferred, or needed activities within nature. It can include active engagement such as physical activities, exercise, gardening, and social activities. It can also include passive engagement such as relaxing, reading, enjoying a meal, or contemplative walking.	X	X
Contact with surrounding life	Observing nature and what is happening in nature; a passive viewing of one's environment.		X
Social opportunities	Socializing with others or socializing with nature itself. It can include active gatherings, meetings, and interacting with others and nature. It can also involve more passively being by oneself, even when with others.	X	X
Culture and connection to the past	Areas that demonstrate human culture, or environmental elements that may spark personal or societal memories.		X
Symbolism/reflection	Spark a connection or symbolism between life and nature; passive observing of nature and seeing it as a metaphor for something in one's own life.		X
Prospect	Well-maintained or groomed green spaces. Can offer opportunities for active or passive engagement as the space itself can be inviting and provide an environment for one to engage the way they need.	X	X
Space	The environment allows for the feeling of calm, tranquility, and rest that can transport one to another world.		X
Rich in species	Refers to the types and numbers of different plants or animals within the ecosystem. Can provide a backdrop for passive engagement of viewing and observing.		x
Sensual pleasures of nature	The active engagement within a natural space, can involve all senses such as seeing, feeling, hearing, smelling, tasting, and moving.	X	
Seasons changing in nature	Identification of changes in nature with each season. This might involve passive engagement such as admiring the changing colors of the leaves, or active engagement such as jumping in a pile of leaves.	X	X
Serene	Areas with calming elements that can provide opportunity for quiet, relaxation, or calm. This can include passive engagement such as listening to a soothing stream.		X
Wild nature	Areas in nature that are not groomed, maintained, or developed.	X	X
Refuge	Areas in nature that offer seclusion, space to be safe and alone.	X	X

Table 1.
Active and passive engagement based on Bengtsson and Grahn list of environmental qualities [17].

Positive health outcomes can be felt from active engagement in nature [1, 49–54]. Bengtsson and Grahn identified qualities of the environment that can offer varying degrees of active and/or passive engagement opportunities. **Table 1** provides an overview of the environmental qualities and identifies whether the qualities align with active or passive engagement [17].

Typically healing gardens are associated with more passive natural experiences, while horticultural therapy is representative of active engagement [55]. Both active and passive engagement can exist within the same space. For instance, healthcare gardens describe garden environments that provide both passive and active engagement opportunities within nature [17]. Grahn's triangle of supporting environments provides a model for understanding active and passive engagement with the environment based on a continuum of well-being [17]. This model posits more passive types of interaction (contemplation, passive experiences of nature) occur when one's sense of well-being is low, and more active and outward-directed engagement (active experience with or in nature) occurs when one's sense of well-being is high. A similar mechanism of action can be seen within sensory modulation wherein the way an individual engages with the environment (the types of activities or stimuli sought) depends on their sensory needs driven by the need for a sense of well-being [41, 56]. The type of active or passive interaction with nature that occurs impacts outcomes [1].

4.2 Dosage

Since 1982, clinicians in Japan have been prescribing time outdoors for health benefits [54]. In Finland, government funded research in 2005 resulted in recommendations of five hours a month being the lowest amount of time needed outdoors to result in health benefits from nature [57]. New Zealand was the first country to begin using the term green prescription to describe healthcare providers prescribing time outdoors to patients. The Green Prescription Active Families survey in 2018 reported 89% of participants who received a green prescription noticed positive health changes, including needing less medication [58]. Healthcare providers in Europe and North America have also been prescribing time outdoors to improve physical and/or mental health [59]. Healthy Parks Healthy People is a movement that started in Australia and is now worldwide that encourages the exploration of parks and nature for health benefits. Rakow and Ibes identified other similar programs including Park Rx, Nature Rx, Walk with a Doc, Campus Nature Rx, and Nature as Medicine that are used to provide exposure to nature as a way to impact physical and mental health positively [59]. Despite all of these global trends towards time spent in nature, less is known about the specific impact that nature exposure has on health given the duration and frequency of nature engagement.

Studies have sought to investigate the relationship between health outcomes (through subjective assessment such as self-report measures or through objective data such as assessment of biomarkers like blood pressure, heart rate, stress hormone level) and the duration of time spent in nature [52, 60, 61]. A duration of 20–30 minutes in nature three times a week has been found to be most effective for providing physiological and psychological benefits [52, 59]. Interestingly, the duration of 20–30 minutes may represent a point of saturation, as additional time spent in nature beyond 30 minutes still has benefits but those resulted at a reduced rate [52]. Smaller increments of 1 and 5-minute green "microbreaks" within a greenspace on campus also resulted in stress relief among college students [60]. Other research suggests that health benefits have been seen from people spending two hours a week in nature, whether time was continuous or accumulated across the week [62]. These benefits were found regardless of personal characteristics such as age, gender, medical conditions, ethnicity, or socio-economic status. White and colleagues also observed that benefits peaked at 200–300 minutes per week with no additional health gains seen after that point [62]. Overall, an adaptive management approach may be beneficial as it allows individuals to change the amount of time and duration spent outdoors while still reaping the benefits of being in nature [52].

4.3 Outcomes related to stress, social isolation, and wellbeing

There is considerable evidence supporting the positive benefits of natural environments on health and well-being, though benefits may vary depending on personal factors such as culture, gender, age, and socioeconomic status and environmental factors such as the location of residence. How each individual receives, and processes sensory information also affects the type of differentiating results. A summary of current evidence outlining the various types of psychological health benefits related to stress, social isolation, and well-being is provided in **Table 2**.

4.4 Physiological outcomes

There are currently limited standardized measures and assessments that reflect assessment of the influence of green spaces on health outcomes [43]. Development in this area is needed to support a robust understanding of the specific features and dosage within outdoor spaces that can result in positive health benefits. Nonetheless, data in the form of biomarkers, self-report measures, observation, or second-hand reporting are utilized to understand the overall health benefits of nature exposure. For example, patients post-surgery in a hospital room with a window view of nature recovered faster and used less pain medication compared to patients post-surgery without window views of nature [94].

While there are variations in research related to dosage (the amount of time spent in nature), the specific type of nature engagement (active or passive), and the exact form that nature takes (wilderness, well-maintained garden, nature view through a window, etc.), current evidence supports many physiological benefits from nature exposure. **Table 3** provides a list of evidence that resulted in physiological benefits. **Figure 3** illustrates the outcomes from the authors' case study.

Outcomes related to stress	Decreased stress	[50, 52, 63–67]
	Decreased anxiety	[53, 54, 68–70]
	Decreased depression	[25, 51, 53, 69, 71–77]
	Decrease in mental distress	[78, 79]
Outcomes related to social isolation	Sense of belonging/positive social interaction	[21, 64, 80–86]
	Greater happiness or life satisfaction	[44, 49, 71, 87–89]
Outcomes related to wellbeing	Emotional wellbeing	[1, 68, 90–93]

Table 2.
Evidence supporting nature exposure health benefits related to stress, social isolation, and wellbeing.

Outcomes related to the nervous system	Increased cognitive functioning	[1, 95]
	Increased memory and attention	[1, 96]
	Improved sleep	[43, 78, 97]
	Improved pain control	[98–100]
Outcomes related to circulatory system	Reduced blood pressure	[101–104]
Global physical outcomes	Improvements post-surgery	[45, 94]

Table 3.
Evidence supporting nature exposure physiological health benefits.

Utilization of Garden by Study Participants.

	Participant 1	Participant 2
Collective Hours Spent in Garden	3–4	5+
Number of Visits to Garden	3–4	5–6
Primary Senses Utilized	Vestibular/ Movement Visual Olfactory	Vestibular/ Movement Visual Olfactory
Top 3 Preferred Sensory Garden Components	Wooden Park Swing Garden Lights Lavender Plant	Wooden Park Swing Wind Spinner Mint Plant

Participants completed a survey that assessed for usage and dosage of the therapeutic sensory garden. They also engaged in individual semi-structured interviews to assess to perceived quality of life themes and understand engagement in the therapeutic garden space. The chart above outlines how each participant utilized the garden, including the number of hours spent in the space and the preferred sensory components they utilized. The results from the interviews indicated that the sensory components, the location of the garden, and time allocated to the garden based on individual schedules/availability were variable components that impacted the utilization of the garden. Student perceptions on quality of life was identified through three themes: 1) connectedness, 2) positive emotional responses and 3) active engagement sensory responses. Participants indicate that quality of life was enhanced by feeling connected when being in the garden. Participants reported emotional responses associated positively with quality of life immediately following participation in the therapeutic sensory garden [30].

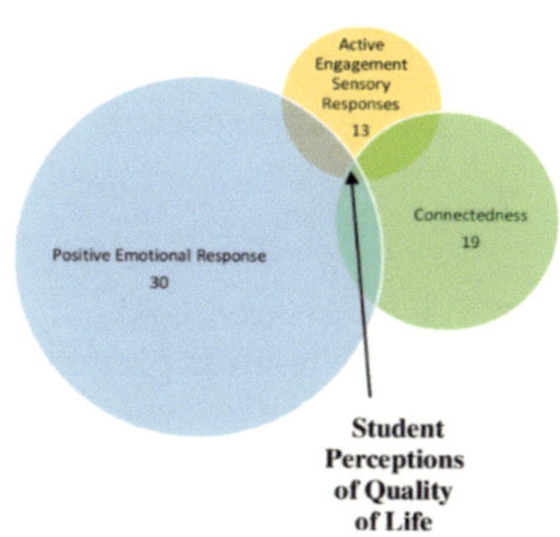

Figure 3.
Case study: outcomes [30].

5. Conclusion

This chapter reviews *what*, *how*, and *why* engagement in green spaces can be therapeutic for well-being. The case study threaded throughout this chapter is representative of how the impact of engagement in green space can be seen in urban environments. While the case study was based on an organic farm, the participants who utilized the space lived in an urban environment. This demonstrates how active or passive engagement in green space is the desired catalyst for physiological and psychological well-being. In any setting, garden spaces that are urban or rural can have positive health benefits. When considering the *what*, *how*, and *why* one must consider the habits and routines of individuals and populations. Nature provides a mechanism for health for individuals, groups, and populations that require a systemic look at how we nurture ourselves and the natural world. Nurturing nature in an urban context can be an opportunity to support health by routinely optimizing nature's benefits. Understanding the science and study of occupational balance resulting from engagement in green spaces and how that impacts individual and population health and well-being outcomes may robustly add to the science and study of urban horticulture.

Acknowledgements

The authors would like to thank the following individuals for their support in their work within active engagement in green space: Tony Miga, farm manager at Eden Hall's campus, for his insight into vegetation, layout, and design to keep a sustainable garden space. Numerous undergraduate and graduate Chatham University students and Eden Hall's farm staff who have supported the sustainable mission of the therapeutic sensory garden at Chatham University's Eden Hall campus. The Chatham University Occupational Therapy program for their support of internal University well-being initiatives. We would also like to thank Chatham University's internal research funding, which has supported the initial construction and longevity of the therapeutic sensory garden, as well as the publication of this chapter in Open Source.

Conflict of interest

The authors have no conflicts of interest to declare.

Notes

It is critical within the review of the literature presented in this chapter and when highlighting the authors' pilot study that the reader know the identities and lived experience of the authors. All three authors are heterosexual, white, female, North Americans from working class families. They have all been trained as occupational therapists and have their doctoral degrees. They work at a University in Northeastern United States. Since the literature reviewed in this chapter bridges both Eastern and Western health traditions, traditional and alternative medicine, it is important to note the assumptions and implicit biases present in the review. The authors primarily utilized peer-reviewed journal articles in English which could limit the scope of knowledge presented. The intent of outlining these characteristics is not to eliminate knowledge, biases, or expectations, but to illuminate potential inaccurate assumptions and acknowledge the presence of how experience and privilege may influence the chapter.

Author details

Kasey Stepansky*, Theresa Delbert and Janet C. Bucey
Chatham University, Pittsburgh, PA, United States of America

*Address all correspondence to: k.stepansky@chatham.edu

References

[1] Bratman GN, Anderson CB, Berman MG, et al. Nature and mental health: An ecosystem service perspective. Science Advances. 2019;**5**:1-14. DOI: 10.1126/sciadv.aax0903 [Epub ahead of print]

[2] Wendelboe-Nelson C, Kelly S, Kennedy M, et al. A scoping review of mapping research on green space and associated mental health benefits. International Journal of Environmental Research and Public Health. 2019;**16**:1-49. DOI: 10.3390/ijerph16122081 [Epub ahead of print]

[3] Masterton W, Carver H, Parkes T, et al. Greenspace interventions for mental health in clinical and non-clinical populations: What works, for whom, and in what circumstances? Health & Place. 2020;**64**:1-19. DOI: 10.1016/j.healthplace.2020.102338 [Epub ahead of print]

[4] National Alliance on Mental Health. Mental Health Benefits of Nature. NAMI California; 2020. Available from: https://namica.org/blog/mental-health-benefits-of-nature/. [Accessed: 27 January 2023]

[5] American Psychological Association. Nurtured by Nature. American Psychological Association; 2020. Available from: https://www.apa.org/monitor/2020/04/nurtured-nature. [Accessed: 27 January 2023]

[6] Harby J. Nature to be Prescribed to GP Patients in Derbyshire. BBC News; 2023. Available from: https://www.bbc.com/news/uk-england-derbyshire-64116268. [Accessed: 27 January 2023]

[7] Miller F. Fighting Nature Deficit Disorder. Newberry Observer; 2022. Available from: https://www.newberryobserver.com/opinion/columns/40236/fighting-nature-deficit-disorder. [Accessed: 27 January 2023]

[8] American Occupational Therapy Association. Occupational therapy practice framework: Domain and process (4th ed.). The American Journal of Occupational Therapy. 2020;**74**:1-87

[9] Yerxa EJ. Some implications of occupational therapy's history for its epistemology, values, and relation to medicine. The American Journal of Occupational Therapy. 1992;**46**:79-83

[10] Wagenfeld A, Atchison B. "Putting the occupation back in occupational therapy:" A survey of occupational therapy practitioners' use of gardening as an intervention. The Open Journal of Occupational Therapy. 2014;**2**:1-19

[11] Wagman P, Håkansson C, Jonsson H. Occupational balance: A scoping review of current research and identified knowledge gaps. Journal of Occupational Science. 2015;**22**:160-169

[12] Liu Y, Zemke R, Liang L, et al. Model of occupational harmony: A chinese perspective on occupational balance. Annals of International Occupational Therapy. 2021;**4**:e228-e235. DOI: 10.3928/24761222-20210601-08 [Epub ahead of print]

[13] Liu CH, Stevens C, Wong SHM, et al. The prevalence and predictors of mental health diagnoses and suicide among U.S. college students: Implications for addressing disparities in service use. Depression and Anxiety. 2019;**36**:8-17

[14] Eriksson T, Westerberg Y, Jonsson H. Experiences of women with stress-related ill health in a therapeutic

gardening program. Canadian Journal of Occupational Therapy. 2011;**78**:273-281

[15] United States Environmental Protection Agency. What is open space/green space? United States Government; 2023. Available from: https://www3.epa.gov/region1/eco/uep/openspace.html

[16] Tillmann S, Tobin D, Avison W, et al. Mental health benefits of interactions with nature in children and teenagers: A systematic review. Journal of Epidemiology and Community Health. 2018;**72**:958-966

[17] Bengtsson A, Grahn P. Outdoor environments in healthcare settings: A quality evaluation tool for use in designing healthcare gardens. Urban Forestry & Urban Greening. 2014;**13**:878-891

[18] Ohly H, White MP, Wheeler BW, et al. Attention restoration theory: A systematic review of the attention restoration potential of exposure to natural environments. Journal of Toxicology and Environmental Health, Part B. 2016;**19**:305-343

[19] Hussein H, Ishak SA, Omar Z. Promotion of inclusive society through therapeutic sensory stimulation garden for the intergenerational society. Environmental Processes Journal. 2016;**1**:161

[20] Association AHT. Definitions and Positions Paper. 2017. Available from: https://www.ahta.org/assets/docs/definitionsandpositionsfinal6.17.pdf. [Accessed: 27 January 2023]

[21] Adevi AA, Mårtensson F. Stress rehabilitation through garden therapy: The garden as a place in the recovery from stress. Urban Forestry & Urban Greening. 2013;**12**:230-237

[22] Jiang S. Therapeutic landscapes and healing gardens: A review of Chinese literature in relation to the studies in western countries. Frontiers of Architectural Research. 2014;**3**:141-153

[23] Chen K, Zhang T, Liu F, et al. How does urban green space impact residents' mental health: A literature review of mediators. International Journal of Environmental Research and Public Health. 2021;**18**:11746

[24] Starry O, Viray A, Park-Egan B, et al. A pilot study considering ecoroofs as therapeutic landscapes. Front Sustain. Cities. 2022;**4**:1-10. DOI: 10.3389/frsc.2022.811306 [Epub ahead of print]

[25] Astell-Burt T, Feng X. Association of urban green space with mental health and general health among adults in Australia. JAMA Network Open. 2019;**2**:e198209

[26] Lee ACK, Maheswaran R. The health benefits of urban green spaces: A review of the evidence. Journal of Public Health. 2011;**33**:212-222

[27] Bettmann JE, Gillis HL, Speelman EA, et al. A meta-analysis of wilderness therapy outcomes for private pay clients. Journal of Child and Family Studies. 2016;**25**:2659-2673

[28] Alcock I, White MP, Lovell R, et al. What accounts for 'England's green and pleasant land'? A panel data analysis of mental health and land cover types in rural England. Landscape and Urban Planning. 2015;**142**:38-46

[29] Skäbäck E, Wadro J, Björk J, et al. The agricultural landscape for recreation. In: Aflakpui G, editor. Agricultural Science. Rijeka, Croatia: Intech; 2012. pp. 225-242

[30] Stepansky K, Delbert T, Bucey JC. Active student engagement within a university's therapeutic sensory garden green space: Pilot study of utilization

and student perceived quality of life. Urban Forestry & Urban Greening. 2022;**67**:127452

[31] UF Health. Horticultural Therapy Worldwide. 2022. Available from: https://wilmotgardens.med.ufl.edu/2022/09/15/horticultural-therapy-worldwide/. [Accessed: 28 January 2023]

[32] American Horticultural Therapy Association. History of Horticultural Therapy. 2023. Available from: https://www.ahta.org/history-of-horticultural-therapy. [Accessed: 27 January 2023]

[33] Gibson S. Horticulture as a therapeutic medium. British Journal of Therapy and Rehabilitation. 1996;**3**:203-209

[34] Owens M, Bunce HLI. Nature-based meditation, rumination and mental wellbeing. International Journal of Environmental Research and Public Health. 2022;**19**:1-19

[35] Ray TN, Franz SA, Jarrett NL, et al. Nature enhanced meditation: Effects on mindfulness, connectedness to nature, and pro-environmental behavior. Environment and Behavior. 2020;**53**:864-890

[36] Unsworth S, Palicki SK, Lustig J. The impact of mindful meditation in nature on self-nature interconnectedness. Mindfulness. 2016;**7**:1052-1060

[37] Furuyashiki A, Tabuchi K, Norikoshi K, et al. A comparative study of the physiological and psychological effects of forest bathing (Shinrin-yoku) on working age people with and without depressive tendencies. Environmental Health and Preventive Medicine. 2019;**24**:46

[38] Yu C-P, Lin C-M, Tsai M-J, et al. Effects of short forest bathing program on autonomic nervous system activity and mood states in middle-aged and elderly individuals. International Journal of Environmental Research and Public Health. 2017;**14**:897

[39] Doimo I, Masiero M, Gatto P. Forest and wellbeing: Bridging medical and forest research for effective forest-based initiatives. Forests. 2020;**11**:791

[40] Bielinis E, Takayama N, Boiko S, et al. The effect of winter forest bathing on psychological relaxation of young Polish adults. Urban Forestry & Urban Greening. 2018;**29**:276-283

[41] Brown A, Tse T, Fortune T. Defining sensory modulation: A review of the concept and a contemporary definition for application by occupational therapists. Scandinavian Journal of Occupational Therapy. 2019;**26**:515-523

[42] Hussein H, Abidin NMNZ, Omar Z. Engaging research and practice in creating for outdoor multi-sensory environments: Facing future challenges. Procedia - Social and Behavioral Sciences. 2013;**105**:536-546

[43] Gonzalez MT, Kirkevold M. Clinical use of sensory gardens and outdoor environments in Norwegian nursing homes: A cross-sectional e-mail survey. Issues in Mental Health Nursing. 2015;**36**:35-43

[44] Van Herzele A, de Vries S. Linking green space to health: A comparative study of two urban neighbourhoods in Ghent, Belgium. Population and Environment. 2012;**34**:171-193

[45] Park S-H, Mattson RH. Effects of flowering and foliage plants in hospital rooms on patients recovering from abdominal surgery. HortTechnology. 2008;**18**:563-568

[46] Kearney AR, Winterbottom D. Nearby nature and long-term care

facility residents: Benefits and design recommendations. The Role of the Outdoors in Residential Environments for Aging. Journal of Housing for the Elderly. 2006;**3-4**:7-28

[47] Kaplan R. The role of nature in the context of the workplace. Landscape and Urban Planning. 1993;**26**:193-201

[48] Ulrich RS, Simons RF, Losito BD, et al. Stress recovery during exposure to natural and urban environments. Journal of Environmental Psychology. 1991;**11**:201-230

[49] Ambrey CL. An investigation into the synergistic wellbeing benefits of greenspace and physical activity: Moving beyond the mean. Urban Forestry & Urban Greening. 2016;**19**:7-12

[50] Antonelli M, Barbieri G, Donelli D. Effects of forest bathing (shinrin-yoku) on levels of cortisol as a stress biomarker: A systematic review and meta-analysis. International Journal of Biometeorology. 2019;**63**:1117-1134

[51] Cohen-Cline H, Turkheimer E, Duncan GE. Access to green space, physical activity and mental health: A twin study. Journal of Epidemiology and Community Health. 2015;**69**:523-529

[52] Hunter MR, Gillespie BW, Chen SY-P. Urban nature experiences reduce stress in the context of daily life based on salivary biomarkers. Frontiers in Psychology. 2019;**10**:1-16. DOI: 10.3389/fpsyg.2019.00722 [Epub ahead of print]

[53] Nutsford D, Pearson AL, Kingham S. An ecological study investigating the association between access to urban green space and mental health. Public Health. 2013;**127**:1005-1011

[54] Song C, Ikei H, Igarashi M, et al. Physiological and psychological effects of a walk in urban parks in fall. International Journal of Environmental Research and Public Health. 2015;**12**: 14216-14228

[55] Marcus CC, Brnes M, editors. Healing Gardens: Therapeutic Benefits and Design Recommendations. Vol. 4. John Wiley and Sons Ltd.; 1999

[56] Fisher AG, Murray EA, Bundy AC. Sensory Integration: Theory and Practice. Philadelphia, PA: Davis Company; 1991

[57] Dose of Nature. Around the World. 2020. Available from: https://www.doseofnature.org.uk/around-the-world1#Sweden. [Accessed: 27 January 2023]

[58] Manatu Hauora Ministry of Health. Green Prescription Active Families Survey Report: May 2018. 2019. Available from: https://www.health.govt.nz/publication/green-prescription-active-families-survey-report-may-2018

[59] Rakow DA, Ibes DC. Campus nature rx: How investing in nature interventions benefits college students. Frontiers in Psychology. 2022;**13**:1-4. DOI: 10.3389/fpsyg.2022.960370 [Epub ahead of print]

[60] Ibes DC, Forestell CA. The role of campus greenspace and meditation on college students' mood disturbance. Journal of American College Health. 2022;**70**:1-8

[61] Meredith GR, Rakow DA, Eldermire ERB, et al. Minimum time dose in nature to positively impact the mental health of college-aged students, and how to measure it: A scoping review. Frontiers in Psychology. 2020;**10**:1-31. DOI: 10.3389/fpsyg.2019.02942 [Epub ahead of print]

[62] White MP, Alcock I, Grellier J, et al. Spending at least 120 minutes a week in

nature is associated with good health and wellbeing. Scientific Reports. 2019;**9**:1-11

[63] Berto R. The role of nature in coping with psycho-physiological stress: A literature review on restorativeness. Behavioral Sciences. 2014;**4**:394-409

[64] Fan Y, Das KV, Chen Q. Neighborhood green, social support, physical activity, and stress: Assessing the cumulative impact. Health & Place. 2011;**17**:1202-1211

[65] Nielsen TS, Hansen KB. Do green areas affect health? Results from a Danish survey on the use of green areas and health indicators. Health & Place. 2007;**13**:839-850

[66] van den Berg AE, Maas J, Verheij RA, et al. Green space as a buffer between stressful life events and health. Social Science & Medicine. 2010;**70**:1203-1210

[67] Ward Thompson C, Aspinall P, Roe J, et al. Mitigating stress and supporting health in deprived urban communities: The importance of green space and the social environment. International Journal of Environmental Research and Public Health. 2016;**13**:440

[68] Bratman GN, Hamilton JP, Hahn KS, et al. Nature experience reduces rumination and subgenual prefrontal cortex activation. Proceedings of the National Academy of Sciences. 2015;**112**:8567-8572

[69] Maas J, Verheij RA, de Vries S, et al. Morbidity is related to a green living environment. Journal of Epidemiology and Community Health. 2009;**63**:967-973

[70] Song C, Joung D, Ikei H, et al. Physiological and psychological effects of walking on young males in urban parks in winter. Journal of Physiological Anthropology. 2013;**32**:18

[71] White MP, Alcock I, Wheeler BW, et al. Would you be happier living in a greener urban area? A fixed-effects analysis of panel data. Psychological Science. 2013;**24**:920-928

[72] Gascon M, Triguero-Mas M, Martínez D, et al. Mental health benefits of long-term exposure to residential green and blue spaces: A systematic review. International Journal of Environmental Research and Public Health. 2015;**12**:4354-4379

[73] Kim W, Lim S-K, Chung E-J, et al. The effect of cognitive behavior therapy-based psychotherapy applied in a forest environment on physiological changes and remission of major depressive disorder. Psychiatry Investigation. 2009;**6**:245

[74] Kondo M, Fluehr J, McKeon T, et al. Urban green space and its impact on human health. International Journal of Environmental Research and Public Health. 2018;**15**:445

[75] McEachan RRC, Prady SL, Smith G, et al. The association between green space and depressive symptoms in pregnant women: moderating roles of socioeconomic status and physical activity. Journal of Epidemiology and Community Health. 2016;**70**:253-259

[76] Sturm R, Cohen D. Proximity to urban parks and mental health. The Journal of Mental Health Policy and Economics. 2014;**17**:19-24

[77] Taylor MS, Wheeler BW, White MP, et al. Research note: Urban street tree density and antidepressant prescription rates—A cross-sectional study in London, UK. Landscape and Urban Planning. 2015;**136**:174-179

[78] Bratman GN, Hamilton JP, Daily GC. The impacts of nature experience on human cognitive function and mental

health. Annals of the New York Academy of Sciences. 2012;**1249**:118-136

[79] Dahlkvist E, Hartig T, Nilsson A, et al. Garden greenery and the health of older people in residential care facilities: A multi-level cross-sectional study. Journal of Advanced Nursing. 2016;**72**:2065-2076

[80] Dadvand P, Bartoll X, Basagaña X, et al. Green spaces and general health: Roles of mental health status, social support, and physical activity. Environment International. 2016;**91**:161-167

[81] de Vries S, van Dillen SME, Groenewegen PP, et al. Streetscape greenery and health: Stress, social cohesion and physical activity as mediators. Social Science & Medicine. 2013;**94**:26-33

[82] Home R, Hunziker M, Bauer N. Psychosocial outcomes as motivations for visiting nearby urban green spaces. Leisure Sciences. 2012;**34**:350-365

[83] Jennings V, Bamkole O. The relationship between social cohesion and urban green space: An avenue for health promotion. International Journal of Environmental Research and Public Health. 2019;**16**:452

[84] Orban E, Sutcliffe R, Dragano N, et al. Residential surrounding greenness, self-rated health and interrelations with aspects of neighborhood environment and social relations. Journal of Urban Health. 2017;**94**:158-169

[85] Piff PK, Dietze P, Feinberg M, et al. Awe, the small self, and prosocial behavior. Journal of Personality and Social Psychology. 2015;**108**:883-899

[86] Sullivan WC, Kuo FE, Depooter SF. The fruit of urban nature. Environment and Behavior. 2004;**36**:678-700

[87] Fleming CM, Manning M, Ambrey CL. Crime, greenspace and life satisfaction: An evaluation of the New Zealand experience. Landscape and Urban Planning. 2016;**149**:1-10

[88] Larson LR, Jennings V, Cloutier SA. Public parks and wellbeing in urban areas of the United States. PLoS One. 2016;**11**:e0153211

[89] MacKerron G, Mourato S. Happiness is greater in natural environments. Global Environmental Change. 2013;**23**:992-1000

[90] Freeman E, Akhurst J, Bannigan K, et al. Benefits of walking and solo experiences in UK wild places. Health Promotion International. 2017;**32**:1048-1056

[91] Hartig T, Mitchell R, de Vries S, et al. Nature and health. Annual Review of Public Health. 2014;**35**:207-228

[92] Hansen MM, Jones R, Tocchini K. Shinrin-yoku (forest bathing) and nature therapy: A state-of-the-art review. International Journal of Environmental Research and Public Health. 2017;**14**:851

[93] Logan AC, Selhub EM. Vis Medicatrix naturae: Does nature 'minister to the mind'? BioPsychoSocial Medicine. 2012;**6**:11

[94] Ulrich RS. View through a window may influence recovery from surgery. Science. 1984;**224**:420-421

[95] Wells NM. At home with nature: Effects of 'greenness' on children's cognitive functioning. Environment and Behavior. 2000;**32**:775-795

[96] Berman MG, Kross E, Krpan KM, et al. Interacting with nature improves cognition and affect for individuals

with depression. Journal of Affective Disorders. 2012;**140**:300-305

[97] Grigsby-Toussaint DS, Turi KN, Williams NJ, et al. Sleep insufficiency and the natural environment: Results from the US behavioral risk factor surveillance system survey. Preventive Medicine. 2015;**78**:78-84

[98] Lechtzin N, Busse AM, Smith MT, et al. A randomized trial of nature scenery and sounds versus urban scenery and sounds to reduce pain in adults undergoing bone marrow aspirate and biopsy. Journal of Alternative and Complementary Medicine. 2010;**16**:965-972

[99] Diette GB, Lechtzin N, Haponik E, et al. Distraction therapy with nature sights and sounds reduces pain during. Chest. 2003;**123**:941-948

[100] Han J-W, Choi H, Jeon Y-H, et al. The effects of forest therapy on coping with chronic widespread pain: Physiological and psychological differences between participants in a forest therapy program and a control group. International Journal of Environmental Research and Public Health. 2016;**13**:255

[101] Duncan M, Clarke N, Birch S, et al. The effect of green exercise on blood pressure, heart rate and mood state in primary school children. International Journal of Environmental Research and Public Health. 2014;**11**:3678-3688

[102] Markevych I, Thiering E, Fuertes E, et al. A cross-sectional analysis of the effects of residential greenness on blood pressure in 10-year old children: Results from the GINIplus and LISAplus studies. BMC Public Health. 2014;**14**:477

[103] Orsega-Smith E, Mowen AJ, Payne LL, et al. The interaction of stress and park use on psycho-physiological health in older adults. Journal of Leisure Research. 2004;**36**:232-256

[104] Shanahan DF, Bush R, Gaston KJ, et al. Health benefits from nature experiences depend on dose. Scientific Reports. 2016;**6**:28551

Chapter 6

Urban Vegetation: Anthropogenic Influences, Public Perceptions, and Wildlife Implications

Catherine Simpson, Cade Coldren, Ioana A. Coman, Caitlyn Cooper-Norris and Gad Perry

Abstract

Urban environments are becoming more common as cities grow and proliferate. Subsequently, their ecosystem services are becoming increasingly more important as climate change impacts urban dwellers, their benefits to human psychological and physical wellbeing are better documented, and biodiversity elsewhere is declining. However, while urban wildlife have received growing attention in recent decades, the importance of urban vegetation has mostly been emphasized in narrow literature niches related to horticulture and landscape architecture. Here, we review literature on multiple uses of urban horticulture, not solely limited to beautification of personal space; the importance of urban vegetation to urban wildlife conservation and sustainability; and how urban vegetation is portrayed in the media.

Keywords: urban vegetation, urban wildlife, classifications, conservation, sustainability

1. Introduction

1.1 Vegetation classifications based on urban uses

Urbanization has caused the global expansion of human-occupied, artificial, or non-natural landscapes in the place of natural, native habitats. While land-use conversion is inevitably associated with the expansion of city boundaries, these novel landscapes can offer beneficial places not just for humans but also for some plants and wildlife. Urban landscapes have many forms, functions, and features which vary based on geographic region and infrastructure. These landscapes consist of many types of vegetation that impact the environment surrounding them and humans in different ways [1]. Plant choice, function, maintenance, and biodiversity create urban landscapes that vary in their sustainability. We begin our discussion of how vegetation is used in urban environments by classifying vegetation types into four primary groups: esthetic/ornamental, functional, and undedicated vegetation.

1.1.1 Aesthetic/ornamental

Over 2000 genera are currently used as ornamental plants [2], defined by their appearance and perceived beauty within a landscape, garden, or interior plant environment and where their primary function is decoration and beautification. Their popularity and use in various landscapes, climates, and gardens throughout the world has risen to make the ornamental plant industry worth over EUR 35.5 billion, a number projected to expand [3]. Ornamental plants are also a fundamental part of human history, having been grown for thousands of years worldwide. However, when it comes to esthetic or ornamental use within an urbanized landscape, diversity in plant selection is limited. A wide range of plants in a garden or landscape may not be practical for homeowners or municipally-managed properties because of cost or availability. Sourcing large numbers of diverse plants is also a limitation, as wholesalers and retail nurseries tend to focus on plants that are popular and profitable [4]. Nurseries also may be limited in plant selection by vendors that only produce certain plants in large quantities. Yet in recent years, a growing number of retail and wholesale stores are offering plants that are more suitable for local environments, climate conditions, and landscape needs. The development of water-wise, native, and specialty landscape plants such as the Texas Superstar® or Proven Winners® plants has further supported interest in species that are better suited to local ecosystems or environments while balancing esthetic appeal and landscape performance.

1.1.1.1 Landscapes

Ornamental plants shape spaces within a landscape and serve to beautify and enhance homes, businesses, and larger, privately-owned areas. In home landscapes, ornamental plants are chosen for their esthetic attributes in relation to a house or building along with their function within the landscape (i.e. shade, color, windbreak). The plants selected by homeowners or landscapers are often chosen based on availability, durability, and cost, rather than ecological or environmental purposes. Landscapes also range in size from large to small and therefore further influence plant selection. However, landscape value and esthetics are subjective. For example, St Hilaire [5] found that landscape preference can be dictated by environment (i.e., desert vs. mesic climates), water shortages, expense of water, and length of time living in a landscape. Because types of landscapes vary widely, such as xeriscape, prairie scape, formal, English, wild, etc., personal preferences account for many of the differences across cities, environments, and between urban regions. Cultural or ethnic background can also account for many landscape preferences. Buijs et al. [6] showed that cultural differences affected landscape preferences in the Netherlands, where immigrants had greater preferences for urban landscapes while native Dutch people gravitated toward wilder settings. In addition to these factors, personal values can also influence plant choices in landscapes. Selection of native plants, pollinator friendly plants, or bird attracting plants has been made easier by the breeding, selection, and marketing of many varieties of landscape plants that cater to these needs. **Figure 1A** illustrates common landscape choices found in Lubbock, Texas, USA, compared to more ecosystem-service centered landscape designs with more native and wildlife friendly plant choices (**Figure 1B-D**). The owners of the yard illustrated in D specifically have pollinators as their management goal. The effort is documented

DOI: http://dx.doi.org/10.5772/intechopen.1001155

Figure 1.
Native vs. non-native urban landscapes. (A) Non-native lawn with few landscape plants, (B) bird friendly plantings, (C) yard with native plants and natural landscaping, and (D) pollinator garden. Photos courtesy of Gad Perry (A and B), Kate LeVering (C) and Catherine Galley (D).

at https://www.inaturalist.org/projects/urban-garden. As of January 14, 2023, there were 734 species documented, including 562 insect species, 56 birds, 48 plants, 38 arachnids, 5 mollusks, 4 mammals, and 2 reptiles. Those responsible for the front (B) and back (C) yards shown are more focused on native, water-wise vegetation and habitats attractive to native and urban species as a whole.

1.1.1.2 Public/botanical gardens

In contrast to private landscapes, botanical gardens are mostly available for the general public to view and enjoy (**Figure 2**). They can be privately sponsored, funded through the government, or supported through grant funding or donations. The primary mission of most botanical gardens is to educate and exhibit plants. More recently, there has been a shift in the education mission of these gardens to include conservation, sustainability education, and public exposure [7]. However, this has had both positive and negative impacts on the organization and operational functionality of these gardens. Many sustainable practices create significant challenges in implementation or financial barriers to adoption [7]. For example, pesticides that help horticulturists suppress pest populations have been phased out because of public perceptions and commitment to sustainable endeavors [7]. Alternatively, increased public adoption of sustainable practices and education initiatives has increased public awareness of "green" practices [7, 8]. Botanical gardens have also spearheaded community and citizen science programs to involve the public in conducting research that contributes to the mission of sustainability as well as understanding responses to climate change [8]. They can also provide networking opportunities for scientists, volunteers, botanists, and ecologists that are conducting projects on similar subjects. Fundamentally, botanical gardens serve a significant role in educating the public about sustainability and creating environmental awareness through science communication efforts.

Figure 2.
The Chicago Botanical Garden located in Chicago, Illinois, USA. Photo courtesy of Catherine Simpson.

1.1.1.3 City beautification

Public greenspaces are often used for beautification and as functional recreational spaces [9]. Green infrastructure components include diverse plant materials, green spaces, green streets, stormwater runoff systems, permeable pavements, green roofs, etc. Green spaces like parks, green streets, and street plantings can positively impact urban ecosystems by reducing temperatures and effects of extreme weather, improving biodiversity and climate resiliency, and mental well-being of humans [10–12]. Thus, green infrastructure is also an important part of urban sustainability [13] . Another concept that can influence city beautification is biophilic design, which emphasizes nature-based systems and design fundamentals to improve sustainability in built environments [14]. Many cities participate in greening and beautification efforts. For example, Keep America Beautiful® is a non-profit organization that partners with communities to build and maintain green spaces (kab.org). To date, they have 695 affiliates across America that participate in clean-up, beautification, and other efforts to educate and bring awareness to environmental concerns. Other communities have taken further steps to improve city beautification by enacting ordinances for landscaping, maintenance, and plant selection for companies that build within city limits. Cities like Austin, Texas, USA, require a Sustainable SITES

Initiative certification for Parks and Recreation Department projects to ensure projects are sustainable [15]. Furthermore, some cities have planting lists that require city-maintained areas to select and plant plants that are non-invasive and will not cause significant damage to surrounding infrastructure [16]. The adoption of such policies may not only improve city beautification, but also biodiversity, plant health, and more widespread urban green space adoption [11, 14, 17, 18].

1.1.1.4 Roadways

While the primary purpose of urban streets is transportation of people and goods, throughways provide space for other public services including exercising, recreation, shopping, and congregating. As such, transportation officials' ideas of proper streetscape composition and management may conflict with city planners' and residents' visions of visually appealing streetscapes. Drivers prefer roadsides with greater amounts of vegetation and increased tree height and density, relative to adjacent commercial properties [19]. Pedestrians and cyclists also select routes based on the amount and type of vegetation present; the presence of large trees and structural diversity in vegetation especially increases preference [20, 21]. Roadside vegetation is correlated with slower driving speeds and reductions in drivers' stress [22]. Roadside vegetation can serve as hazards, but vehicle crashes into vegetation are less common in urban areas compared to rural areas due to lower speed limits and more clearly defined boundaries of urban roads [22, 23].

1.2 Functional

While esthetics can be considered a function of plants on its own, urban plants can also serve other purposes to the broader environment. Among these uses are ecological or conservation, food production, recreation, and roadways/streetscapes.

1.2.1 Ecological or conservation

Though ecological and conservation functions may not be the original intended function of urban plants, urban vegetation provides critical ecosystem services including cooling, heat mitigation, air filtration, storm water retention and filtration, and habitat and food sources for urban wildlife. Urban vegetation contributes to human health and well-being through therapeutic and calming effects, outdoor recreation, and reducing exposure to environmental stressors (e.g., ultraviolet radiation, heat exposure, air and water pollution). These benefits are discussed in detail in the "Quality of Life" subsection of this chapter. Due to increased recognition of ecosystem services, ecological and conservation purposes have grown in popularity as intended functions of urban vegetation. Some specific examples of urban vegetation established for ecological and conservation functions include rain gardens, pollinator gardens, green roofs, and xeriscaping.

1.2.2 Food production

Community gardens differ from private gardens, landscapes, and largescale food production systems in being accessible by members of the public and serving as areas for food production in urban or peri-urban locations [24]. Yet, community gardens vary widely in their purpose and function. These purposes vary from bringing

Figure 3.
Heart of Lubbock Community Garden located in Lubbock, Texas, USA. Photo courtesy of Jonah Trevino.

community members together to garden; promoting healthy practices; engaging children; promoting food security (**Figure 3**), land conservation or restoration; or allowing users to connect with nature. Community gardens can help people access more diverse foods, which aids in social justice causes and reduces undernutrition caused by calorie dense, nutrient poor diets [25]. Furthermore, Alaimo et al. [26], Carney et al. [27], and Litt et al. [28] showed that adults who participated in community gardens significantly increased vegetable and fruit intake. They can also serve to address conservation and sustainability on a community level [9]. Additionally, the ecosystem services that community gardens can provide are not limited to humans alone. They provide habitats for pollinators, beneficial insects, wildlife, and benefit the environment as well. A recent meta-analysis shows that the quantity of crops (ex. 129–200% yield increases per m^2) grown in urban settings can be substantial [29, 30]. Another approach gaining ground in the food production sector is urban vertical farming (e.g., [31]). Typically using hydroponic or aeroponic techniques, these high-tech efforts offer the promise of producing food close to urban consumers. Using reduced footprints and inputs, including reduced water usage, increased yields per m^2, and reduced food miles, they can be quite profitable [32]. However, they still only produce a narrow variety of foods and require advanced technologies that may not be suitable for conditions in many developing countries [33].

1.2.3 Recreational

While some may consider community gardens to be recreational space, these gardens are generally limited to people who enjoy gardening as an activity. However, recreational uses of plants are not limited only to gardens. Recreational greenspaces can include parks, sports fields, playgrounds, public seating areas, lots, and other areas where nature can be experienced [34]. These functional spaces are maintained to fulfill particular needs within society, such as exercise, etc. The plants selected for the spaces

Figure 4.
Recreational spaces in Lubbock, Texas, USA. (A) Play area for children located in a park, (B) park that has been minimally maintained and has led to prairie dog repopulation, (C) walking trail with limited accessibility for handicapped persons, and (D) grey fox seen near a building on Texas Tech University campus. Photos A–C courtesy of Gad Perry, photo D courtesy of Emily Stamm.

that are designated for sports typically are less diverse than those in greenbelts or natural areas. Grass species dominate sports fields and athletic areas [1] and are more aggressively maintained than non-sports areas. However, functional spaces may offer a large, planted area for wildlife and other species to inhabit (**Figure 4**). For example, an extensive review by Petrosillo et al. [35] found that while golf courses may negatively impact water and soil environments, they increase biodiversity and ecosystem services, particularly in amphibian, bird, turtle, and bee species. These findings are further supported by Tanner and Gange [36] and Guzy et al. [37] who observed increased biodiversity and species richness near golf courses in urban areas. Other areas that are designated as recreational may include urban remnant prairies or forests and refer to patches of space cities use as a natural park but contain minimal built structures. This increases the likelihood of ecological plant and animal diversity within these spaces, but also provides opportunities for species to encroach into urbanized areas where they may not be welcome (**Figure 4B** and **D**).

1.3 Undedicated vegetation

1.3.1 Lots

Though these areas do not have a dedicated function, they provide ecosystem services for urban humans and animals that would not otherwise be provided if the space was filled with buildings or impervious surfaces. In an early example, Darlington [38] summarized the flora associated with urban walls, primarily in the United Kingdom, and the animals that rely on them and help them disperse. Vacant or abandoned areas comprise about 15% of the land area in cities [39]. These areas receive little to no maintenance and are dominated by spontaneous vegetation comprised of a variety of species. Plants occurring in these areas may include remnant species purposefully planted when the area had a designated function, in addition to species that were unintentionally introduced through a variety of abiotic (i.e., wind, water) or biotic (i.e., endozoochory, epizoochory by animals) dispersal mechanisms. The vegetative composition and structure in vacant areas is highly influenced by biophysical and social factors. In a study of vacant lots in Chicago, for example, species richness was most affected by the amount of trash in the lot, home ownership around the lot, proportion of the lot covered by turfgrass, and the proportion of area around the lot covered by built structures [40].

2. Applications of vegetation in urban environments

Vegetation functionality changes in different urban environments. For example, vegetation can serve as pollutant filters, erosion control, stormwater buffers, as well as spaces for wildlife habitats or human recreation. Often this relies upon the selection of plants that have particular roles in the ecosystem. Selection of plants for environmental or water conservation, carbon sequestration, pollutant mitigation, wildlife attractants, or the improvement of quality of life are some of the primary roles that plants can play in an urban landscape.

2.1 Sustainable and conservation practices

2.1.1 Native plants

Native plant species have grown in popularity for urban gardens due to a variety of reasons such as being adapted to local climate conditions and providing critical habitat and ecosystem services [41–43]. Additionally, native species are increasingly adopted to replace the large amounts of non-native species that have come to characterize urban vegetation; between 30% and 50% of urban plant species are non-native [44–46]. Several cities have adopted ordinances mandating that native plants be used in and around new commercial developments and public landscapes and promoting management practices that make urban vegetation more hospitable to wildlife and conducive to natural ecosystem functions [47–49]. Additional details about the effects of urban vegetation on wildlife are provided elsewhere in this chapter, particularly in the sections titled "Importance of Vegetation for Urban Wildlife" and "Case Study." Native plantings vary in scale; homeowners and business owners may dedicate flowerbeds or entire lawn areas to native plants [50, 51]. In contrast to typical turfgrass lawns, native plantings are typically managed as low- to moderate-intensity, "near-natural" states with little to no fertilizer or irrigation inputs (**Figure 1B** and **C**).

2.1.2 Water conservation and management

Garden and lawn irrigation is estimated to comprise approximately 30% of residential urban water use in the United States, totaling nearly 30 billion liters per day [52]. Outdoor water use may be as high as 60% in arid climates [52]. Sustainable gardening encourages the use of water conservation and adoption of species that require less water. "Xeriscaping", "water-smart", and "water-wise" landscaping are similar concepts with the common end goal of attractive water-efficient landscapes [53, 54]. The inclusion of low water-using plants that are tolerant of the regional climate is a key component, particularly in light of climate change challenges [55].

As mentioned above, adaptation to local climate and soils is one of the drivers for adopting native vegetation. Species selection should balance water usage with water infiltration. A study by Chang et al. [54] reported that xeriscaping, although water use-efficient, had less infiltration and greater runoff than turf lawns. Ideally species selected for water-efficient landscaping will require little irrigation while maintaining moderate to high infiltration rates and low runoff.

2.2 Services plants provide

2.2.1 Climate and pollution mitigation

Due to their large amounts of impermeable surfaces, cities are especially vulnerable to extreme precipitation flooding events. Impervious surfaces result in low infiltration and high runoff; subsequently, the potential for flooding and runoff containing large quantities of chemical and physical hazards increases [56]. Vegetated areas provide critically important permeable surfaces within the largely impervious urban environment. The replacement of impervious surfaces by vegetation slows the speed and reduces the total amount of runoff. Zabret and Šraj [57] calculated that reducing asphalt surfaces in a parking lot by 35.5% through tree planting would result in a 7–17% reduction in runoff. Additionally, vegetated areas assist in stormwater treatment and improve water quality through bioretention processes [58]. Green infrastructure specifically used for stormwater retention and treatment includes rain gardens, constructed wetlands, bioretention basins, roadside swales, filter strips, and green roofs [56, 58, 59]. Chinese researchers proposed the "sponge city" concept in 2013 as a systematic method to retain, slow, and clean water in urban areas [60]. In 2015, 16 cities were selected to pilot the program; the program expanded to another 14 cities in 2016. Goals included retaining and reusing 70% of urban stormwater by 2020 and reusing up to 80% of stormwater by the 2030s. Though the infrastructure and concepts mentioned in the previous sentences are specifically deployed to capture and treat stormwater, vegetation in residential yards and private properties can serve similar functions, though they may be restrained. Vegetable and herb gardens are not typically thought of in terms of their ability to retain stormwater or improve runoff water quality. However, Whittinghill et al. [61] reported that vegetable-producing green roofs had similar water retention capabilities and reduced NO_3^- concentrations in runoff as green roofs with *Sedum* spp., a succulent ground cover commonly used on green roofs.

Urban vegetation improves air quality by modifying deposition and dispersion of gaseous molecules and particulate matter. Because plants have a high surface area to volume ratio, deposition on vegetative material is enhanced relative to smooth, synthetic surfaces [62]. Vegetative species with hairy, thin, and/or waxy leaves have increased deposition relative to species without these characteristics [63]. Urban vegetation alters wind direction and speed, therefore affecting the dispersion pattern and dilution of pollutants. In a review of urban vegetation effects on deposition and dispersion, Janhäll [64] recommended vegetation be shorter and/or closer to surfaces to allow dilution of emissions by clean air. They also recommended vegetation be planted close to the source of pollutants to increase deposition, such as planting vegetative barriers near roadways. Lastly, because polluted air passing above vegetation is not filtered, vegetative barriers should be porous enough for air to pass through and deposit pollutants, but solid enough to encourage air to pass close to the surface.

2.2.2 Quality of life

The health, well-being, and quality of life benefits of vegetation are extensive. Urban vegetation plays a critical role in urban residents' well-being. Integrating biophilic design principles can reduce energy consumption, lower the urban heat island effect, and improve human resilience in the face of climate change [11, 14]. Integration of these concepts can also make humans more aware of nature and, as a result, encourage them to care more about protecting natural spaces [65]. Vegetated spaces provide the public with cultural ecosystem services, as greenspaces contribute to increases in property value, tourism, improving population mental and physical health, as well as supporting social cohesion and interactions [10, 66–68]. Urban green spaces have also been linked to reduction of violent crimes, improvement of cognitive development, reduction of noise levels, and other social benefits [69–73]. These benefits also translate to specialized green spaces such as university campuses, where McFarland et al., Speake et al., Van Den Bogerd et al., and Trevino et al. [74–77], among others, have shown that green spaces can improve perceptions of the campus, academic performance, and mental health. Frumkin et al. [78] reviewed many studies on the health benefits of nature contact and reported better sleep, improved mental health, life satisfaction, social connections, healing, birth outcomes, child development, pain control, immune function, and general health as well as reduced aggression, stress, blood pressure, heart failure, obesity, diabetes, asthma, and mortality. These observations have led to the use of plants in complementary therapies such as horticultural therapy (registered horticultural therapist directed horticultural activities with the aim to achieve specific goals within an established plan or treatment) or therapeutic horticulture (use of horticultural activities as a therapeutic modality facilitated by registered horticultural therapists or trained horticulturalists) to improve quality of life for different populations [79]. Therapeutic horticulture can involve different modalities, generally active or passive interactions with plants. When plants are used as an active tool in therapies, improvements to physical and mental health have been seen. This can be particularly useful in vulnerable populations such as children, elderly, military veterans, and the mentally ill because of the lack of significant, adverse impacts [80–87]. In a study by Fleming [88], U.S. military veterans reported that 'veteran to farmers' programs that utilized interactions with plants helped alleviate symptoms of Post Traumatic Stress Disorders (PTSD) while giving participants a source of meaningful work and skills. These findings have been echoed all over the world and throughout history, from Zen gardens, to nature assisted therapy programs established after World War I and II [82]. This further supports the theory that nature and vegetation appeals to humans on a fundamental level and improves our quality of life.

2.2.3 Wildlife and invertebrate habitats

Vegetation specifically utilized for ecological and conservation functions is the most conducive to providing habitat for wildlife and pollinators. However, various studies have documented the benefits of other vegetation classes and functional types on urban vertebrates and invertebrates. Recreational spaces, botanical gardens, ornamental plants, roadway vegetation, community and private gardens, and undedicated vegetation have all been shown to have positive effects on urban wildlife and invertebrates [89–92]. The degree of positive effects and potential negative effects vary with the specific function and plant species utilized.

Diverse urban plant assemblages with a greater abundance of plants and high species richness do not always equate to suitable quality habitat for urban wildlife. Urbanized environments support increased non-native plant species abundance and richness [93]. This increase in non-native plant species is primarily driven by human-mediated trade and transport, whether intentional or unintentional [94]. Effects of alien vegetation on urban wildlife and invertebrates range from positive, to negative, to negligible [43, 95–97]. Generally, native plant species benefit urban fauna with benefits (e.g., greater abundance, diversity, occupancy, and richness) to native animal species attributed more frequently to native than exotic plant species [43]. Non-native animals tend to benefit more from introduced plant species. Wildlife responses to plants' "nativeness" are quite complex though. Requirements for habitat and food resources vary considerably across species, and urban vegetation's delivery of necessary resources is ultimately more important than plant origin [98, 99]. The success of wildlife in cities depends on species' ability to acclimate to suboptimal resources in the urban environment. Additionally, even if an urban space contains primarily native plant species, native wildlife abundance and richness may be low due to competition with nonnative biota which are better adapted to the urban surroundings. Relationships between urban vegetation and urban wildlife are discussed in detail under "Importance of Vegetation for Urban Wildlife" below.

2.3 Challenges

2.3.1 Availability or access

While the benefits of green spaces in urban areas are extensive, a major challenge regarding urban vegetation and green spaces is equitable access and availability for all communities. In urban centers where availability of green space is already scarce, preferential distribution has been shown to benefit more affluent communities [17]. Low-income communities have more limited access to green spaces, or lower quality and less maintained green spaces [67]. This may contribute to socioeconomic inequality and inequity and can result in poorer health and psychological impacts as well. Moreover, city planners and horticulturists must be aware that inequitable green space construction can further marginalize low-income populations and communities of color. There have been efforts to combat this by installing more green spaces in communities where there are none, but this may lead to gentrification, resident displacement, and risk further marginalizing these communities [17]. To avoid "ecological or green gentrification," developers and city officials must consider the impact of improvements on property values and focus environmental cleanup and green development according to community needs rather than applying generic approaches [17, 18]. These approaches can include community gardens, restoration, or small-scale sites that are "just green enough" to provide community benefits without displacing the residents they were originally meant to serve [17]. Additionally, better access to green spaces can narrow the magnitude of differences between socioeconomic classes [67]. Access can also refer to accommodation of disabled populations. Health inequality is an issue of concern in most urban centers, and by addressing access to green spaces and risks associated with poorly maintained spaces, overall public health can be improved. Fortunately, laws regarding accommodations for disabled persons such as wheelchair access, smooth pathways, etc. have become more common. However, not all public greenspaces are equipped with these modifications due to infrastructure, expense, and labor limitations. This further supports the need of city officials to include adaptations to improve access to green spaces to include a more diverse population with variable needs.

2.3.2 Adoption and acceptance

Alternative garden and yard designs and adoption of sustainable practices (e.g., using native plants, reduced fertilizer, reduced water, locally-sourced materials) may require fewer external inputs and be more ecologically friendly [55]. However, these landscapes may be less attractive to traditional landowners than conventional yards, due to perceived increases in maintenance [100] and personal environmental attitudes [101]. Additionally, adoption of alternative designs and sustainable practices may have high costs up front [102, 103], and it may take months to years to see the benefits [104] whereas homeowners expect fairly immediate and obvious benefits. If results are slow or ambiguous, people lose interest and support for the practice or species diminishes. Although people report wanting to increase sustainability and the associated benefits of vegetated areas in cities, many are less willing to personally adopt certain practices [105, 106]. Other studies have reported that residents are willing to adopt sustainable practices and alternative garden and yard designs if the perceived benefits outweigh the costs and any downsides associated. For example, Helfand et al. [107] reported that consumers were willing to pay more for esthetically pleasing prairie gardens which provided more ecological benefits than traditional turf lawns, despite increases in costs associated with the native plantings. Turf replacement participation is significantly correlated with replacement rebate rates, owner-occupancy, median income, and residents' environmental attitudes [101, 108]. Residents who adopt environmentally friendly landscapes can increase adoption in their neighborhoods; Pincetl et al. [108] reported that 36% of properties with turf replacement had nearby properties with partially- or fully-replaced landscapes. Personal preferences may deem some species and maintenance practices as suitable for a resident's personal yard, but unsuitable for public green spaces, and vice versa. For example, dense vegetation around residents' personal property may provide them with a sense of privacy, but dense vegetation in public spaces such as parks may decrease residents' sense of safety, especially at night [109, 110].

2.3.3 Invasiveness and escape

Non-native plant species are typically more tolerant of the altered microclimates, hydrology, and soil structure that characterize urban areas. Non-native plants also experience less pressure from competing vegetation and natural enemies (i.e., pests, herbivores) in urban environments than natural areas [93]. These characteristics encourage the spread of introduced species from areas where they were originally planted. This spread is not limited to the neighboring urban surroundings; urban vegetation often escapes into outlying rural areas, where it can have negative impacts on natural vegetation and ecosystem processes.

2.3.4 Negative attributes

There are some concerns that urban green spaces may exacerbate urban residents' allergies and rhinitis. There are several studies that have shown both improvements in allergies and negative impacts on allergies [78, 111]. In contrast to desired objectives, urban vegetation can act as a source of pollutants and chemicals into the urban environment [56]. Despite being highly pervious, vegetated areas may be sources of nutrients, herbicides, insecticides, sediments, and fecal bacteria. Fertilizers and pesticides applied to vegetation may runoff or leach out. Since these spaces often provide recreational opportunities and habitat, deposition of pet and wildlife feces commonly occurs. Vegetated sites may also serve as secondary sources of contaminants; resuspension or leaching may release pollutants originating from other sources [56, 112]. For this reason, it is essential to determine the pollutant loads and/or concentrations that the structure will receive and the percent reduction to be achieved prior to selecting the type of stormwater control measures or green infrastructure to be used and how it will be arranged in the urban drainage system [113].

Urban vegetation may intercept gaseous pollutants and particulate matter to a high degree, and vegetation grown on polluted sites may also take in contaminants through their roots. In turn, urban residents may be exposed to these pollutants when interacting with urban vegetation. Particulate matter may be deposited on edible plant parts, and/or the particulate matter deposited on foliage or taken up through the roots may be translocated to fruits and vegetables [114, 115]. Adverse health effects can occur through ingestion, inhalation, or skin contact with contaminated foods, soil, or non-food plant matter [78, 116]. Because we tend to focus on the health-positive aspects of vegetation in urban environments, people may not consider the unintended, potentially harmful effects on their health from urban vegetation.

3. Importance of vegetation for urban wildlife

Awareness of the conservation potential of urban wildlife goes back several decades [117, 118], with some mentions emerging even earlier (see [119], pp. xi–xii; 2005). Early work has given way to increasingly complex and nuanced research on urban ecology. Yet conservation-oriented studies of urban vegetation have often not touched on wildlife to any great extent [120, 121] and vice versa [122, 123]. A separate thread, reviewed elsewhere in this chapter, has focused on urban vegetation as a design component concentrating primarily on esthetics and human wellbeing [124] . Until recently, relatively little of this literature emphasized the importance of using native species. A third theme, the most recent, has emerged as awareness of the impacts of global climate change on urban areas has grown and concentrates on the value that urban vegetation can offer in increasing shade, taking up CO_2 and other pollutants, etc. [125]. This discussion, too, has mostly occurred independently of the other two, with ecosystem services being considered separately from biodiversity and from human health and esthetic pleasure. Instead, we advocate a One Health-based approach [126]; for an urban application see [127] that jointly addresses the wellbeing of wildlife, humans, and the environment (**Figure 5**).

Ever since the pioneering work of MacArthur and MacArthur [128], ecologists have been explicitly aware that animal diversity depends on the diversity of plant-created habitats. Wildlife biologists have adopted an "if you build it they will come" attitude toward supporting the populations of species they considered desirable via intentional manipulation of environmental vegetation structure. "In a landscape planting scheme for wildlife, it is also important to consider vegetation" ([119] p. 102). This is doubly true in urban environments, which by definition provide a structurally simplified habitat with limited and unfamiliar characteristics.

What do we know about the importance of vegetation structure for urban wildlife? In a series of studies on urban gardens in the United Kingdom, Smith et al. [129]

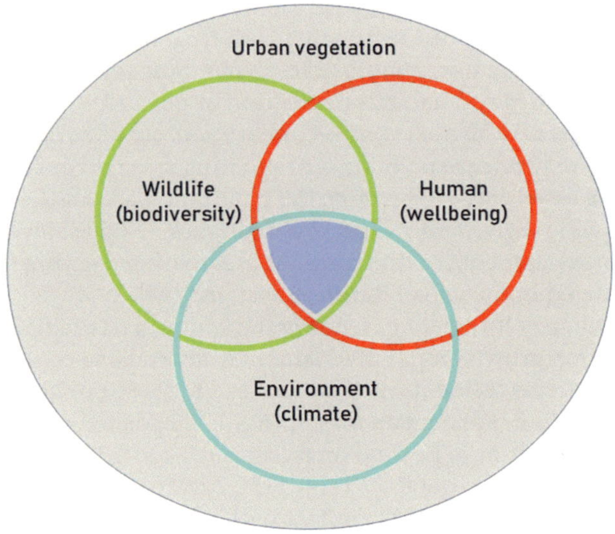

Figure 5.
One Health-based approach to urban vegetation, wildlife, environment, and humans. Figure courtesy of Gad Perry.

documented urban vegetation and the variation among sites, found that most of the plant diversity was comprised of non-native species, and concluded that "garden floras may offer many more resources to wildlife than is implied by their species composition" [129]. In a separate study from the Central Arizona–Phoenix Long-Term Ecological Research site in the United States, Lerman and Warren [130] more explicitly connected urban plant and bird diversities. They found that that the abundance of native desert birds increased as native plant landscaping was emplaced and that racial and economic identity had important bearing on resident access to this biodiversity. The greater animal biodiversity associated with native vegetation appears to be a broad pattern, though non-native plant species also support substantial animal biodiversity [43].

Studying related issues in Baltimore, Maryland, USA, Rega-Brodsky et al., [131] surveyed the plant and bird communities in 150 vacant lots and found that the lots supported biodiversity but did not meet the esthetic preferences of residents, who preferred lots with more trees that they perceived as "well-maintained." This type of vegetation, termed Urban Spontaneous Vegetation (USV) by Riley et al. [132], sustains urban wildlife and provides residents with a connection with nature and human physical and mental health benefits. Rega-Brodsky et al. [131] suggested that planting additional trees in vacant lots can simultaneously address the inclinations of neighbors while improving habitat for urban birds. However, some of the wildlife supported by USVs is considered objectionable, whether for esthetic reasons or their ability to vector human diseases [132].

Birds and pollinators remain the most studied taxa in this context [43], but there are some studies on other vertebrates. Although vegetation and its structural importance do not appear in the titles of any of the 40 chapters or 13 case studies presented in Mitchell et al. [133], there are some reptile and mammal studies [43]. For example, in a lizard study from Hobart, Tasmania, Australia, Jellinek et al. [134] found that reptile communities were influenced by vegetation type and structure. Similarly, while residential gardens in Australia support native mammals and have conservation value, the vegetative structure and diversity and their possible influences were not studied by Van Helden et al. [135, 136]. Likewise, although urban green spaces in Accra, Ghana, supported some mammal diversity, the importance of specific vegetative traits was not explored [137]. However, studies of mammals in forest patches in Durban, South Africa, showed that mammal diversity was positively affected by vegetative structure in urban forest remnants [138, 139]. The recent review by Brum et al. [140] did not indicate that the importance of vegetation structure has been of great interest in the herpetological literature, a definite research gap, and the situation appears to be similar in the urban mammal literature.

The relationship between vegetation structure and animal diversity is not limited to vertebrates and pollinators. In a multi-taxa study focusing on urban forest patch size in Basel, Switzerland, Melliger et al. [141] found that arthropod diversity depended on trophic rank and habitat specificity, with vegetative structural diversity predictive of ant and spider diversity. Plants species richness decreased with urbanization, but even small urban forests held substantial arthropod biodiversity [141]. More recently, Mills et al. [142] showed that more vegetatively diverse urban green spaces also had greater soil microbial diversity. Similarly, Francoeur et al. [143] reported that urban lawns with greater vertical complexity in Montreal, Canada, had greater arthropod biodiversity.

Finally, it is important to remember that the relationship between plants and animals is bi-directional. While urban plants provide essential physical structure and food for existence and sustenance of wildlife, the ecosystem services that wildlife provides for urban plants cannot be ignored either. Perhaps most obvious is the role of pollinators. Cities hold a greater diversity of pollinating insects than do nearby agricultural areas, especially in areas where hardscapes are uncommon [144, 145]. In return, those insects help fertilize urban plants and ensure their propagation, a service particularly important for urban farms. No less important, urban birds provide essential dispersal services for fruiting plants [146], just as adding artificial perches can attract birds and enhance plant dispersal in restoration projects [147].

4. Lubbock, Texas, USA: a case study

Lubbock, Texas, USA (**Figure 6**), is a medium-sized city, with a population of approximately 261,000 [148], that is located on the southern Great Plains of North America (33.5779 °N, 101.8552 °W). The climate of Lubbock is semi-arid, receiving an average annual rainfall of 48.6 cm, snowfall of 22.9 cm, and with a mean annual temperature of 15.8°C [149].

Lubbock provides a unique setting for evaluating the impacts of urban vegetation on wildlife species. Except for native riparian vegetation found on the eastern edge of town, most of the urbanization is occurring into areas of either native rangeland or agricultural fields, most of which are planted primarily in cotton (*Gossypium hirsutum*). Few trees are found in the native rangeland, with most being honey mesquite (*Prosopis glandulosa*). Scattered throughout Lubbock are a series of small, modified playa lakes, predominantly with permanent water and serving as part of a runoff control system that also has a role in public recreation. The dominant trees found naturally along playa edges are cottonwood (*Populus deltoides*) and black willow (*Salix nigra*). Thus, the majority of trees and shrubs found in town are a result of plantings as urbanization has proceeded over time. **Table 1** lists the trees and shrubs commonly planted in parks and residential areas in the rapidly urbanizing portions of

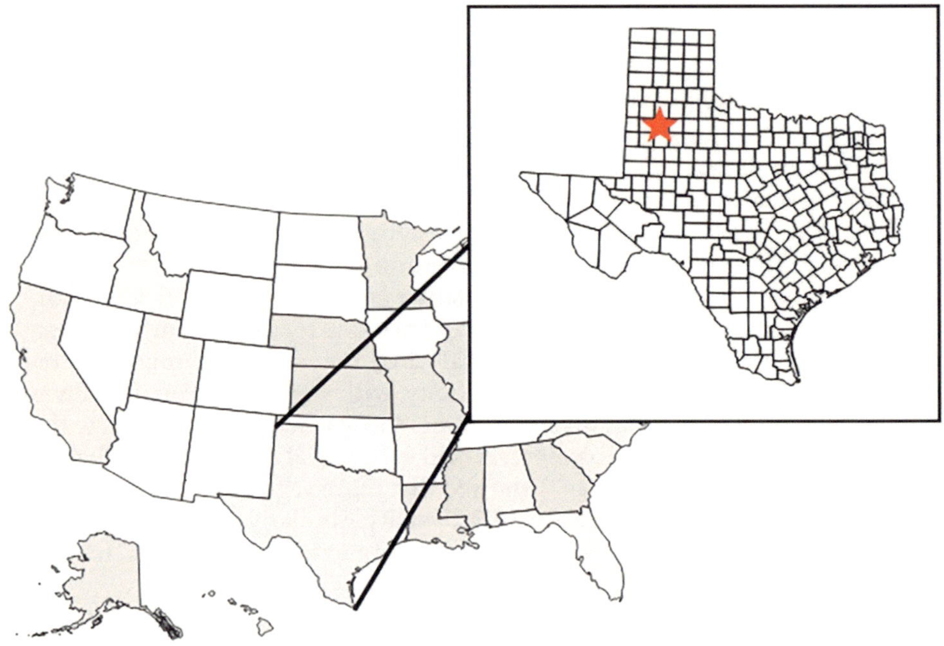

Figure 6.
Map of the USA, Texas, and Lubbock.

Lubbock (to the west, the south, and the southwest). Note that these contain a mix of native and introduced species. Many of the species listed as native to North America were not historically found in this region.

Several studies have been completed in recent years on the wildlife in Lubbock, all of which included some aspect of the impact of vegetation on animal distribution, diversity, richness, or population density. Organismal groups included amphibians [150], birds (Coldren unpublished data, and Freeman [151]), and mammals [152]. **Table 2** lists the major species evaluated in each group, which again include a mix of native and introduced taxa, as well as some that are native to North America but not originally found in this region.

Ramesh [150] found that the most important features explaining the presence of amphibians at lakes in the urban zone were the type and structure of vegetation adjacent to lakes, as well as the land use matrix and age of development surrounding the lakes. Lakes with amphibian detections tended to be in areas of more recent development and supported greater cover of native emergent and fringing vegetation. This vegetation provided pollutant filtration, oviposition sites, and refuge from predators. Lakes in older parts of town were located in parks more heavily managed and manicured (**Figure 7**), and they supported fewer amphibians. In these parks, emergent and fringing vegetation were constantly being removed and the surrounding turfgrass was regularly mowed, reducing their suitability for amphibians. Additionally, the land use matrix surrounding lakes in older developed areas contained more roads, more impervious cover, and less vegetated areas. This resulted in reduced rates of movement between lakes, effectively eliminating the ability of amphibians to recolonize lakes from which they had disappeared.

Two recent studies have been conducted on birds in Lubbock. Freeman [151] evaluated the effects of human population density, canopy cover, neighborhood age,

Common name	Scientific name	Native status
American elm	*Ulmus americana*	Native
Ash	*Fraxinus* sp.	Some species native, some introduced
Bald cypress	*Taxodium distichum*	Native
Black walnut	*Juglans nigra*	Native
Black willow	*Salix nigra*	Native
Cedar elm	*Ulmus crassifolia*	Native
Chinese Elm	*Ulmus parvifolia*	Introduced
Chinese pistache	*Pistacia chinensis*	Introduced
Cottonwood	*Populus deltoides*	Native
Crepemyrtle	*Lagerstroemia* sp.	Introduced
Desert willow	*Chilopsis linearis*	Native
Golden rain tree	*Koelreuteria paniculata*	Introduced
Honey locust	*Gleditsia triacanthos*	Native
Honey mesquite	*Prosopis glandulosa*	Native
Juniper	*Juniperus* sp.	Native
Lilac	*Syringa* sp.	Introduced
Live oak	*Quercus virginiana*	Native
Maple	*Acer* sp.	Native
Mulberry	*Morus* sp.	Some species native, some introduced
Pear	*Pyrus* sp.	Introduced
Pecan	*Carya illinoinensis*	Native
Persimmon	*Diospyros* sp.	Some species native, some introduced
Pine	*Pinus* sp.	Native
Red oak	*Quercus* sp.	Native
Sophora	*Sophora* sp.	Native
Sugar hackberry	*Celtis laevigata*	Native
Sumac	*Rhus* sp.	Native
Sweet gum	*Liquidambar styraciflua*	Native
Sycamore	*Platanus occidentalis*	Native
White oaks	*Quercus* sp.	Native
Yaupon	*Ilex vomitoria*	Native

Table 1.
Trees and shrubs commonly planted in parks and residential areas in Lubbock, Texas, USA.

and distance to the city center in relation to the distribution and richness of three groups of birds. *Exploiter* species such as rock pigeons and sparrows are often non-native and adept at exploiting human-controlled resources. *Adapter* species such

Common name	Scientific name	Native status
Amphibians		
American Bullfrog	*Lithobates catesbeiana*	Native to North America but not Texas
Great Plains Narrow-Mouthed Toad	*Gastrophryne olivacea*	Native
Plains Spadefoot	*Spea bombifrons*	Native
Spotted Chorus Frog	*Pseudacris clarkii*	Native
Texas Toad	*Anaxyrus speciosus*	Native
Birds		
American Robin	*Turdus migratorius*	Native
Barn Swallow	*Hirundo rustica*	Native
Blue Jay	*Cyanocitta cristata*	Native
Chimney Swift	*Chaetura pelagica*	Native
Curve-billed Thrasher	*Toxostoma curvirostre*	Native
Eurasian Collared-Dove	*Streptopelia decaocto*	Introduced
European Starling	*Sturnus vulgaris*	Introduced
Great-tailed Grackle	*Quiscalus mexicanus*	Native
House Finch	*Haemorhous mexicanus*	Native
House Sparrow	*Passer domesticus*	Introduced
Killdeer	*Charadrius vociferus*	Native
Mississippi Kite	*Ictinia mississippiensis*	Native
Mourning Dove	*Zenaida macroura*	Native
Northern Cardinal	*Cardinalis cardinalis*	Native
Northern Mockingbird	*Mimus polyglottos*	Native
Red-winged Blackbird	*Agelaius phoeniceus*	Native
Rock Pigeon	*Columba livia*	Introduced
Western Kingbird	*Tyrannus verticalis*	Native
White-winged Dove	*Zenaida asiatica*	Native to Texas but not Lubbock
Mammals		
Coyote	*Canis latrans*	Native
Domestic Dog	*Canis familiaris*	Introduced
Feral Cat	*Felis catus*	Introduced
Gray Fox	*Urocyon cinereoargenteus*	Native
Raccoon	*Procyon lotor*	Native
Red Fox	*Vulpes vulpes*	Native to North America but not Texas
Virginia Opossum	*Didelphis virginiana*	Native

Table 2.
Focal wildlife species studied in Lubbock, Texas by Carter [152], Coldren (unpublished data), Freeman [151], and Ramesh [150].

Figure 7.
Typical vegetative structure in an older park in Lubbock, Texas, USA. Photo courtesy of Cade Coldren.

as great-tailed grackles and American robins are native birds that have adapted to human-dominated landscapes. *Uncommon* species such as the curve-billed thrasher are native species that are not well adapted to urban or suburban settings. The best predictor of exploiter species presence was neighborhood age, which also served as a surrogate for a diverse vegetative structure, since plantings mature over time. Human population density and neighborhood age were found to be the best predictors for adapter species. Canopy cover was also found to be a good predictor of adapter species as it was generally accompanied by greater vertical stratification of vegetation, and therefore supported a greater diversity of birds. The uncommon species were found on the periphery of town, but also in areas where remnant native vegetation occurred. Additionally, socioeconomic and educational factors were found to play a role, as higher education and greater disposable income tended to result in greater vegetative diversity, and thus greater bird diversity and richness. Overall, it appears that decisions by individual homeowners regarding plantings in gardens and yards have a greater impact on birds than city planners do.

In unpublished data, Coldren investigated the distribution and density of native and non-native bird species in 20 parks across the urbanizing areas of Lubbock. Vegetative and land use data included in the analyses were tree diversity, mean tree canopy cover, herbaceous cover, development within a park, water, and several metrics for a 250 m buffer around each park, including impervious cover, green space, and extent of residential, commercial, and industrial areas. Cluster analysis revealed three distinct clusters based on differences in vegetative characteristics. The first cluster included the older parks with more mature trees and highest tree diversity and mean tree canopy cover (**Figure 7**). These parks tended to be the closest to the center of town. The second cluster was intermediate in terms of tree diversity and mean tree canopy cover, and these parks tended to be found at a greater distance from the center of town. The last cluster were the youngest parks, with the lowest tree diversity and mean tree canopy cover. These parks were the farthest from the center of town and were also in areas with the highest commercial development and greatest amount of impervious cover. Avian diversity declined from the first cluster to the second, and was lowest in the third, although these differences were not statistically significant. Interestingly, density of individual bird species tended to peak in the park cluster with intermediate tree diversity and mean canopy cover, while it was lowest in the cluster of youngest parks.

Carter [152] installed cameras throughout Lubbock to capture occurrences of mammals and evaluated various metrics of vegetation and land use within 200 m of each camera to assess which factors appeared more important in understanding mammalian distributions across the urban zone. Vegetation and land use metrics included canopy cover, green space cover, manicured grass, bare soil, anthropogenic cover, water, total road length, and the distance to the nearest 4-lane road. Carter [152] found that most of the species captured on camera were not associated with any of the habitat variables under consideration, although there were exceptions. Coyotes were seen only on the periphery of town, in areas with some intact native rangeland vegetation. Raccoons tended to occur in areas with greater canopy cover, as well as surface water and patches of wetlands. Gray foxes were associated with lower canopy cover in summer, but not the rest of the year. Feral cats had a positive association with green space cover in autumn, but negative associations with all vegetative variables in winter. Additionally, feral cat distribution and density appeared to be tied to food sources, whether supplied directly by humans or indirectly through the occurrence of prey species such as rodents and small birds, which tended to be found in areas with greater vegetative diversity and structure.

Considering the three organismal groups, it appears that wildlife use of urban areas of Lubbock is more strongly based on the age, structure, and diversity of vegetation, rather than the presence or absence of individual plant species (**Table 1**). The wildlife species found in these studies (**Table 2**) tend to be generalists and are not known to be reliant on one or a small suite of plant species. Vegetative age and structure tend to be a result of diverse plantings, including trees, shrubs, and herbaceous plants, and the time necessary for them to become mature plants. This appears to be the strongest requirement for the greatest diversity and use of urban vegetation by wildlife in this city.

5. Urban plants and their portrayal in the media

Traditionally, urban vegetation is not often covered by the media. Compared to animals, where more negative or unusual occurrences between wild animals and humans lead to media coverage spikes [153], urban plants do not seem that newsworthy. However, in the past years, especially with COVID-19 lockdowns, climate change and crisis focus and solution journalism, increasingly more news stories about plants emerged. These same topics also gained traction in public spheres and public online spaces. For example, on platforms like TikTok conservation experts are trying to educate the public, promote conservation and strategic gardening, and offer advice on what people can do to bring or save native vegetation in their own spaces.

While it is well documented that news media impact public opinions and can further impact actual policies [154, 155]. The related literature is also scarce—more recently a few studies looked at plant pest invasion coverage in the news and social media [156, 157]; or urban plant foraging attitudes [158].

Generally, plants covered in the media fall into the following categories:

1. *Domestic*—domestic plants, such as house plants, flower shops/flower markets, flowers in parks, gardening, edible plants etc.

2. *Wild*—wild plants such wild vegetation in the city, or wild parks, or forgotten poorly maintained areas.

3. *Useful/utility*—how and why plants can be used or help people (e.g., recipes, cures, prevention medicine, but also helping mental health, or general human health, etc.).

4. *Dangerous/damaging*—how and why plants can hurt humans or their environment (e.g., poisonous, allergies, financial damage such as trees falling over houses/cars, deaths, etc.).

5. *Gray area*—species such as cannabis that are somewhat controversial.

We conducted a non-exhaustive analysis of 100 news stories published in 2022. These news stories were mostly U.S. based and were extracted from repeated searches in news aggregators such as Google News, Yahoo, etc. Before analysis, we verified that each story indeed focused on urban related vegetation. Besides ensuring this criterion, stories were not further selected or checked for content until the actual analysis.

Our findings show some positive trends and some concerning ones. The majority of the analyzed news stories focused on wild urban vegetation and the usefulness or utility of vegetation (see **Figure 8**).

Furthermore, compared to wildlife, where our previous analysis [153] showed media vilified animals and emphasized their dangerous traits, in the context of

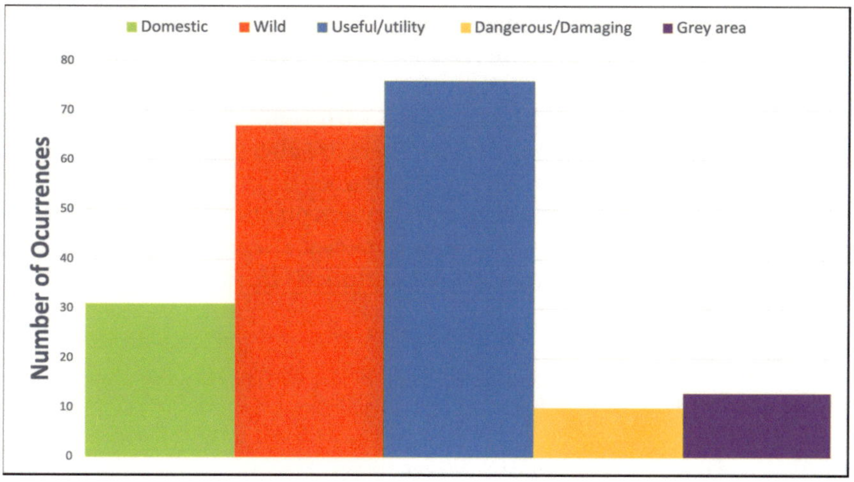

Figure 8.
A breakdown of 100 popular media news stories about wild urban vegetation. These categories were not mutually exclusive, and some news stories contained more than one angle, so they appear more than once.

vegetation the majority of the news stories focused on the positive traits of plants and how they can be useful. These were diverse and solution- or how-to stories about topics such as:

- The need to integrate more native plants in cities, lawns, and gardens;
- The case for/need for controlling or sometimes even using wild invasive plant species that thrive in challenging environments;
- Urban revegetation and positive effects on humans; going foraging in the city; etc.

Often these stories were almost militant or activist about the high need to integrate wild plants and more general wildlife in urban areas in order to mitigate the negative effects of the climate crisis. We found fewer stories that were focused toward the negative or damaging attributes of wild plants, about topics such as invasive species displacing native plants, the need to pay attention when foraging as some plants are not edible or could even cause severe health damages or death, etc. Finally, compared to our previous wildlife analysis [153], in this current case, it is also encouraging that the majority of these news stories about wild urban vegetation used experts as sources. Experts were quoted, paraphrased, and usually given a good amount of news space. Taken together these findings suggest there is an increasing interest in positively covering such topics especially through solution journalism or how-to news stories and offer a platform for conservation and vegetation experts to share their knowledge and tips. Scientists should thus take advantage of these opportunities and continue to better educate the public in regard to how to best use, protect, and revive wild urban vegetation.

6. Conclusions

In conclusion, urban vegetation is important for humans and wildlife. Not only does urban vegetation provide ecosystem services such as wildlife habitats, climate mitigation, and more, but it also promotes human health, beautification, and psychological benefits. Plants are generally viewed as positive contributions to an environment, but many factors can affect this perception. Non-native species, water usage, and adverse associations with safety or health factors can negatively affect the functionality or services plants provide. There are also issues regarding accessibility, gentrification, and greenwashing that detract from beneficial aspects of plants. Yet, overall, the positive benefits and services plants provide in urban areas are widespread and affect humans and animal species. Conservation and education accomplishments have come a long way but continued efforts will ensure progress is made on these fronts.

Conflict of interest

The authors declare no conflict of interest.

Author details

Catherine Simpson*, Cade Coldren, Ioana A. Coman, Caitlyn Cooper-Norris and Gad Perry
Texas Tech University, Lubbock, Texas, USA

*Address all correspondence to: catherine.simpson@ttu.edu

References

[1] Fuller RA, Irvine K. Interactions between people and nature in urban environments. In: Urban Ecology. Cambridge: Cambridge University Press; 2010. pp. 134-171

[2] Chen J, Chen J. Ornamental plant research inaugural editorial. Ornamental Plant Research. 2021;**1**:1-2

[3] Gabellini S, Scaramuzzi S. Evolving consumption trends, marketing strategies, and governance settings in ornamental horticulture: A grey literature review. Horticulturae. **8**. Epub ahead of print 2022. DOI:10.3390/horticulturae8030234

[4] Torres-Camacho KA, Meléndez-Ackerman EJ, Díaz E. Intrinsic and extrinsic drivers of yard vegetation in urban residential areas: Implications for conservation planning. Urban Ecosystem. 2017;**20**:403-413

[5] St. Hilaire R, VanLeeuwen DM, Torres P. Landscape preferences and water conservation choices of residents in a high desert environment. HortTechnology. 2010;**20**:308-314

[6] Buijs AE, Elands BHM, Langers F. No wilderness for immigrants: Cultural differences in images of nature and landscape preferences. Landscape and Urban Planning. 2009;**91**:113-123

[7] Johnson V. Identity, sustainability, and local setting at U.S. Botanical Gardens. Organization & Environment. 2012;**25**:259-285

[8] Primack RB, Ellwood ER, Gallinat AS, et al. The growing and vital role of botanical gardens in climate change research. The New Phytologist. 2021;**231**:917-932

[9] Tharrey M, Perignon M, Scheromm P, et al. Does participating in community gardens promote sustainable lifestyles in urban settings? Design and protocol of the JArDinS study. BMC Public Health. 2019;**19**:1-10

[10] Fuller RA, Irvine KN, Devine-Wright P, et al. Psychological benefits of greenspace increase with biodiversity. Biology Letters. 2007;**3**:390-394

[11] Beatley T, Newman P. Biophilic cities are sustainable, resilient cities. Sustain. 2013;**5**:3328-3345

[12] Beyer KMM, Kaltenbach A, Szabo A, et al. Exposure to Neighborhood green space and mental health: Evidence from the survey of the health of Wisconsin. International Journal of Environmental Research and Public Health. 2014;**11**:3453-3472

[13] Im J. Green streets to serve urban sustainability: Benefits and typology. Sustain. **11**. Epub ahead of print 1 November 2019. DOI:10.3390/SU11226483

[14] Africa J, Heerwagen J, Loftness V, et al. Biophilic design and climate change: Performance parameters for health. Frontiers in Built Environment. 2019;**5**:28

[15] City of Austin Improves Resilience and Community Health Through Landscape Development in Parks and Green Spaces | AustinTexas.gov. Available from: https://www.austintexas.gov/news/city-austin-improves-resilience-and-community-health-through-landscape-development-parks-and-green-spaces [Accessed: 20 January 2023]

[16] Street Tree Planting Approved Species List : NYC Parks. Available from: https://www.nycgovparks.org/

trees/street-tree-planting/species-list [Accessed: 20 January 2023]

[17] Wolch JR, Byrne J, Newell JP. Urban green space, public health, and environmental justice: The challenge of making cities 'just green enough'. Landscape and Urban Planning. 2014;**125**:234-244

[18] Newman A. Inclusive urban ecological restoration in Toronto, Canada. Human Dimensions of Ecological Restoration. Washington, DC: Island Press; 2011. pp. 63-75. DOI: 10.5822/978-1-61091-039-2

[19] Wolf KL. Freeway roadside management: The urban forest beyond the white line. Journal of Arboriculture. 2003;**29**:127-136

[20] Van Dongen RP, Timmermans HJP. Preference for different urban greenscape designs: A choice experiment using virtual environments. Urban Forestry & Urban Greening. 2019;**44**:126435

[21] Nawrath M, Kowarik I, Fischer LK. The influence of green streets on cycling behavior in European cities. Landscape and Urban Planning. 2019;**190**:103598

[22] Van Treese J II, Koeser A, Fitzpatrick G, et al. A review of the impact of roadway vegetation on drivers' health and well-being and the risks associated with single-vehicle crashes. Arboricultural Journal. 2017;**39**:179-193

[23] Harvey C, Aultman-Hall L. Urban streetscape design and crash severity. Transportation Research Record. 2015;**2500**:1-8

[24] Ferris J, Norman C, Sempik J. People, land and sustainability: Community gardens and the social dimension of sustainable development. Social Policy and Administration. 2001;**35**:559-568

[25] Bamji MS, Murty PVVS, Sudhir PD. Nutritionally sensitive agriculture—An approach to reducing hidden hunger. European Journal of Clinical Nutrition. 2021;**75**:1001-1009

[26] Alaimo K, Packnett E, Miles RA, et al. Fruit and vegetable intake among urban community gardeners. Journal of Nutrition Education and Behavior. 2008;**40**:94-101

[27] Carney PA, Hamada JL, Rdesinski R, et al. Impact of a community gardening project on vegetable intake, food security and family relationships: A community-based participatory research study. Journal of Community Health. 2012;**37**:874-881

[28] Litt JS, Soobader MJ, Turbin MS, et al. The influence of social involvement, neighborhood aesthetics, and community garden participation on fruit and vegetable consumption. American Journal of Public Health. 2011;**101**:1466

[29] Payen FT, Evans DL, Falagán N, et al. How much food can we grow in urban areas? Food production and crop yields of urban agriculture: A meta-analysis. Earth's Futur. 2022;**10**:2022EF002748

[30] Touliatos D, Dodd IC, Mcainsh M. Vertical farming increases lettuce yield per unit area compared to conventional horizontal hydroponics. Food and Energy Security. 2016;**5**:184-191

[31] Kalantari F, Tahir OM, Joni RA, et al. Opportunities and challenges in sustainability of vertical farming: A review. Journal of Landscape Ecology. 2018;**11**:35-60

[32] Avgoustaki DD, Xydis G. Indoor vertical farming in the urban nexus context: Business growth and resource savings. Sustainability. 2020;**12**:1965

[33] Van Delden SH, SharathKumar M, Butturini M, Graamans LJA, et al. Current status and future challenges in implementing and upscaling vertical farming systems. Nature Food. 2021;**2**:944-956

[34] What is Open Space/Green Space? | Urban Environmental Program in New England | US EPA. Available from: https://www3.epa.gov/region1/eco/uep/openspace.html [Accessed: 14 January 2023]

[35] Petrosillo I, Valente D, Pasimeni MR, et al. Can a golf course support biodiversity and ecosystem services? The landscape context matter. Landscape Ecology. 2019;**34**:2213-2228

[36] Tanner RA, Gange AC. Effects of golf courses on local biodiversity. Landscape and Urban Planning. 2005;**71**:137-146

[37] Guzy JC, Price SJ, Dorcas ME. The spatial configuration of greenspace affects semi-aquatic turtle occupancy and species richness in a suburban landscape. Landscape and Urban Planning. 2013;**117**:46-56

[38] Darlington A. Ecology of Walls. London: Heinemann Educational Publishers; 1981

[39] Branas CC, South E, Kondo MC, et al. Citywide cluster randomized trial to restore blighted vacant land and its effects on violence, crime, and fear. In: Proceedings of the National Academy of Sciences. 2018. pp. 2946-2951

[40] Anderson E, Minor E. Assessing social and biophysical drivers of spontaneous plant diversity and structure in urban vacant lots. Science of the Total Environment. 2019;**653**:1272-1281

[41] Crewe K. Arizona native plants and the urban challenge. Landscape Journal. 2013;**32**:215-229

[42] Stevenson PC, Bidartondo MI, Blackhall-Miles R, et al. The state of the world's urban ecosystems: What can we learn from trees, fungi, and bees? Plants, People, Planet. 2020;**2**:482-498

[43] Berthon K, Thomas F, Bekessy S. The role of 'nativeness' in urban greening to support animal biodiversity. Landscape and Urban Planning. 2021;**205**:103959

[44] Lososova Z, Chytry M, Danihelka J, et al. Biotic homogenization of urban floras by alien species: The role of species turnover and richness differences. Journal of Vegetation Science. 2016;**27**:452-459

[45] Heywood VH. The nature and composition of urban plant diversity in the Mediterranean. Flora Mediterranea. 2017;**27**:195-220

[46] Gaertner M, Wilson JRU, Cadotte MW, et al. Non-native species in urban environments: Patterns, processes, impacts and challenges. Biological Invasions. 2017;**19**:3461-3469

[47] Weschester County. County executive George Latimer signs executive order to protect native plants in Westchester. 2018. Available from: *www.westchestergov.com*, https://www.westchestergov.com/home/all-press-releases/7783-county-executive-george-latimer-signs-executive-order-to-protect-native-plants-in-westchester [Accessed: 11 April 2022]

[48] Talal ML, Santelmann MV. Vegetation management for urban park visitors: A mixed methods approach in Portland, Oregon. Ecological Applications. 2020;**30**:2079

[49] Town of Victor Conservation Board. Town of Victor, Native Plant Manual. New York: Victor; 2018

[50] Johnston MR, Balster NJ, Zhu J. Impact of residential prairie gardens

on the physical properties of urban soil in Madison, Wisconsin. Journal of Environmental Quality. 2016;**45**:45-52

[51] Fukase J, Simons AM. Increased pollinator activity in urban gardens with more native flora. Applied Ecology and Environmental Research. 2016;**14**:297-310

[52] U.S. Environmental Protection Agency. Water-smart landscapes: Start with WaterSense. 2021

[53] Brelsford C, Abbott JK. How smart are 'water smart landscapes'? Journal of Environmental Economics and Management. 2021;**106**:102402

[54] Chang B, Wherley B, Aitkenhead-Peterson JA, et al. Effects of residential landscape composition on surface runoff generation. Science of the Total Environment. 2021;**783**:146977

[55] Alizadeh B, Hitchmough J. A review of urban landscape adaptation to the challenge of climate change. International Journal of Climate Change Strategies and Management. 2019;**11**:178-194

[56] Muller A, Osterlund H, Marsalek J, et al. The pollution conveyed by urban runoff: A review of sources. Science of the Total Environment. 2020;**709**:136125

[57] Zabret K, Šraj M. Can urban trees reduce the impact of climate change on storm runoff? Urbani Izzivi. 2015;**26**:165-178

[58] Liu A, Egodawatta P, Goonetilleke A. Ranking three water sensitive urban design (WSUD) practices based on hydraulic and water quality treatment performance: Implications for effective stormwater treatment design. Water. 2022;**14**:1296

[59] Ekka SA, Rujner H, Leonhardt G, et al. Next generation swale design for stormwater runoff treatment: A comprehensive approach. Journal of Environmental Management. 2021;**279**:111756

[60] Chan FKS, Griffiths JA, Higgitt D, et al. "Sponge City" in China—A breakthrough of planning and flood risk management in the urban context. Land Use Policy. 2018;**76**:772-778

[61] Whittinghill LJ, Rowe DB, Andresen JA, et al. Comparison of stormwater runoff from sedum, native prairie, and vegetable producing green roofs. Urban Ecosystem. 2014;**18**:13-29

[62] Diener A, Mudu P. How can vegetation protect us from air pollution? A critical review on green spaces' mitigation abilities for air-borne particles from a public health perspective - with implications for urban planning. Science of the Total Environment. 2021;**796**:148605

[63] Saebo A, Popek R, Nawrot B, et al. Plant species differences in particulate matter accumulation on leaf surfaces. Science of the Total Environment. 2012;**427-428**:347-354

[64] Janhäll S. Review on urban vegetation and particle air pollution—Deposition and dispersion. Atmospheric Environment. 2015;**105**:130-137

[65] Chen G, Sun W. The role of botanical gardens in scientific research, conservation, and citizen science. Plant Diversity. 2018;**40**:181-188

[66] Barton J, Pretty J. What is the best dose of nature and green exercise for improving mental health—A multi-study analysis. Environmental Science & Technology. 2010;**44**:3947-3955

[67] Braubach M, Egorov A, Mudu P, et al. Effects of Urban Green Space on Environmental Health, Equity and Resilience. In: Kabisch N, editor. Nature-based Solutions to Climate Change Adaptation in Urban Areas, Theory and Practice of Urban Sustainability Transitions. pp. 187-205

[68] Daniel TC, Muhar A, Arnberger A, et al. Contributions of cultural services to the ecosystem services agenda. Proceedings of the National Academy of Sciences of the United States of America. 2012;**109**:8812-8819

[69] Shepley M, Sachs N, Sadatsafavi H, et al. The impact of green space on violent crime in urban environments: An evidence synthesis. International Journal of Environmental Research and Public Health. 2019;**16**:4-10

[70] Zhou X, Parves RM. Social benefits of urban green space. Management of Environmental Quality: An International Journal. 2012;**23**:173-189

[71] Dadvand P, Nieuwenhuijsen MJ, Esnaola M, et al. Green spaces and cognitive development in primary schoolchildren. Proceedings of the National Academy of Sciences of the United States of America. 2015;**112**:7937-7942

[72] Cohen P, Potchter O, Schnell I. The impact of an urban park on air pollution and noise levels in the Mediterranean city of Tel-Aviv, Israel. Environmental Pollution. 2014;**195**:73-83

[73] Klingberg J, Broberg M, Strandberg B, et al. Influence of urban vegetation on air pollution and noise exposure—A case study in Gothenburg, Sweden. Science of the Total Environment. 2017;**599-600**: 1728-1739

[74] McFarland AL, Waliczek TM, Zajicek JM. The relationship between student use of campus green spaces and perceptions of quality of life. HortTechnology. 2008;**18**:232-238

[75] Speake J, Edmondson S, Nawaz H. Everyday encounters with nature: Students' perceptions and use of university campus green spaces. Human Geography. 2013;**7**:21-31

[76] Van Den Bogerd N, Coosje Dijkstra S, Seidell JC, et al. Greenery in the university environment: Students' preferences and perceived restoration likelihood. PLoS One. 2018;**13**:1-19

[77] Trevino JE, Monsur M, Lindquist CS, et al. Student and nature interactions and their impact on mental health during the COVID-19 pandemic. International Journal of Environmental Research and Public Health. **19**. Epub ahead of print 2022. DOI:10.3390/ijerph19095030

[78] Frumkin H, Bratman GN, Breslow SJ, et al. Nature contact and human health: A research agenda. Environmental Health Perspectives. **125**. Epub ahead of print 1 July 2017. DOI:10.1289/EHP1663

[79] American Horticultural Therapy Association. Definitions and Positions. 2023. Available from: https://www.ahta.org/ahta-definitions-and-positions

[80] Stowell DR, Owens GP, Burnett A. A pilot horticultural therapy program serving veterans with mental health issues: Feasibility and outcomes. Complementary Therapies in Clinical Practice. DOI:S174438811830166X

[81] Kirk A, Karpf A, Carman J. Therapeutic garden design and veterans affairs: Preparing for future needs. Journal of Therapeutic Horticulture. 2010;**20**:66-76

[82] Poulsen DV, Stigsdotter UK, Refshage AD. Whatever happened to the soldiers? Nature-assisted therapies for veterans diagnosed with post-traumatic stress disorder: A literature review. Urban Forestry & Urban Greening. 2015;**14**:438-445

[83] Kamioka H, Tsutani K, Yamada M, et al. Effectiveness of horticultural therapy: A systematic review of randomized controlled trials. Complementary Therapies in Medicine. 2014;**22**:930-943

[84] Determan J, Craig Gaulden Davis Mary Anne Akers F, Albright T, et al. Impact of Biophilic Learning Spaces on Student Success | 1 The Impact of Biophilic Learning Spaces on Student Success. 2019. Available from: https://cgdarch.com/wp-content/uploads/2019/12/The-Impact-of-Biophilic-Learning-Spaces- [Accessed: 15 February 2022]

[85] Han KT. Influence of passive versus active interaction with indoor plants on the restoration, behaviour and knowledge of students at a junior high school in Taiwan. Indoor and Built Environment. 2018;**27**:818-830

[86] Lu S, Wu F, Wang Z, et al. Evaluation system and application of plants in healing landscape for the elderly. Urban Forestry & Urban Greening. 2021;**58**:126969

[87] Park SY, Song JS, Kim HD, et al. Effects of interior plantscapes on indoor environments and stress level of high school students. Journal of the Japanese Society for Horticultural Science. 2008;**77**:447-454

[88] Fleming LL. Veteran to farmer programs: An emerging nature-based programming trend. Journal of Therapeutic Horticulture. 2015;**25**:27-48

[89] Gallo T, Fidino M, Lehrer EW, et al. Mammal diversity and metacommunity dynamics in urban green spaces: Implications for urban wildlife conservation. Ecological Applications. 2017;**27**:2330-2341

[90] Wood EM, Esaian S. The importance of street trees to urban avifauna. Ecological Applications. 2020;**30**:2149

[91] Felderhoff J, Gathof AK, Buchholz S, et al. Vegetation complexity and nesting resource availability predict bee diversity and functional traits in community gardens. Ecological Applications:e2759

[92] Prudic KL, Cruz TMP, Winzer JI, et al. Botanical gardens are local hotspots for urban butterflies in arid environments. Insects. 2022;**13**:865

[93] Cadotte MW, Yasui SLE, Livingstone S, et al. Are urban systems beneficial, detrimental, or indifferent for biological invasion? Biological Invasions. 2017;**19**:3489-3503

[94] Essl F, Lenzner B, Bacher S, et al. Drivers of future alien species impacts: An expert-based assessment. Global Change Biology. 2020;**26**:4880-4893

[95] Matteson KC, GA L. Small scale additions of native plants fail to increase beneficial insect richness in urban gardens. Insect Conservation and Diversity. 2011;**4**:89-98

[96] Gray ER, van Heezik Y. Exotic trees can sustain native birds in urban woodlands. Urban Ecosystem. 2016;**19**:315-329

[97] Mata L, Andersen AN, Morán-Ordóñez A, et al. Indigenous plants promote insect biodiversity in urban greenspaces. Ecological Applications. 2021;**31**:2309

[98] Sander HA, McCurdy JD. Urban vegetation and songbird nesting guilds: Relationships and implications for conservation and management. Urban Forestry & Urban Greening. 2021;**64**:127308

[99] Duren AM, Williams CK, D'amico V. Microhabitat factors associated with occupancy of songbirds in suburban forest fragments in the eastern United States. The American Midland Naturalist. 2017;**178**:189-202

[100] Khachatryan H, Rihn A, Hansen G, et al. Landscape aesthetics and maintenance perceptions: Assessing the relationship between homeowners' visual attention and landscape care knowledge. Land Use Policy. 2020;**95**:104645

[101] Zhang X, Khachatryan H. Interactive effects of homeowners' environmental concerns and rebate incentives on preferences for low-input residential landscapes. Urban Forestry & Urban Greening. 2021;**65**:127322

[102] van der Meulen SH. Costs and benefits of green roof types for cities and building owners. Journal of Sustainable Development of Energy, Water and Environment Systems. 2019;**7**:57-71

[103] Drosou N, Soetanto R, Hermawan F, et al. Key factors influencing wider adoption of blue–green infrastructure in developing cities. Water. 2019;**11**:1234

[104] Langner AN, Manu A, Nath DA. Dynamic soil properties across a suburban landscape. Soil Science Society of America Journal. 2013;**77**:1284-1293

[105] Watson CJ, Carignan-Guillemette L, Turcotte C, et al. Ecological and economic benefits of low-intensity urban lawn management. Journal of Applied Ecology. 2020;**57**:436-446

[106] Avolio M, Pataki DE, Jenerette GD, et al. Urban plant diversity in Los Angeles, California: Species and functional type turnover in cultivated landscapes. Plants, People, Planet. 2020;**2**:144-156

[107] Helfand GE, Park JS, Nassauer JI, et al. The economics of native plants in residential landscape designs. Landscape and Urban Planning. 2006;**78**:229-240

[108] Pincetl S, Gillespie TW, Pataki DE, et al. Evaluating the effects of turf-replacement programs in Los Angeles. Landscape and Urban Planning. 2019;**185**:210-221

[109] Lis A, Iwankowski P. Why is dense vegetation in city parks unpopular? The mediative role of sense of privacy and safety. Urban Forestry & Urban Greening. 2021;**59**:126988

[110] Rahm J, Sternudd J, Johansson M. "In the evening, I don't walk in the park": The interplay between street lighting and greenery in perceived safety. Urban Design International. 2021;**26**:42-52

[111] Fuertes E, Markevych I, von Berg A, et al. Greenness and allergies: Evidence of differential associations in two areas in Germany. Journal of Epidemiology and Community Health. 2014;**68**:787-790

[112] Flanagan K, Branchu P, Boudahmane L, et al. Retention and transport processes of particulate and dissolved micropollutants in stormwater biofilters treating road runoff. Science of the Total Environment. 2019;**656**:1178-1190

[113] Behroozi A, Niksokhan M, Nazariha M. Developing a simulation-optimisation model for quantitative and qualitative control of urban run-off using best management practices. Journal of Flood Risk Management. 2018;**11**:5340-5351

[114] Xiong T-T, Leveque T, Austruy A, et al. Foliar uptake and metal(loid) bioaccessibility in vegetables exposed to particulate matter. Environmental Geochemistry and Health. 2014;**36**:897-909

[115] Przybysz A, Stepniak A, Malecka-Przybysz M, et al. Particulate matter accumulation on apples and plums: Roads do not represent the greatest threat. Agronomy. 2020;**10**:1709

[116] Jim CY, Chen WY. Assessing the ecosystem service of air pollutant removal by urban trees in Guangzhou (China). Journal of Environmental Management. 2008;**88**:665-676

[117] Adams LW, Leedy DL, editors. Integrating man and nature in the metropolitan environment. In: Proceedings of a National Symposium on Urban Wildlife. Columbia, Maryland: National Institute for Urban Wildlife; 1986. pp. 4-7

[118] Stenberg K, Shaw WW, editors. Wildlife Conservation and New Residential Developments. Tucson, Arizona: University of Arizona Press; 1986

[119] Adams LW. Urban Wildlife Habitats: A Landscape Perspective. Minneapolis: U of Minnesota Press; 1994

[120] Bastin L, Thomas CD. The distribution of plant species in urban vegetation fragments. Landscape Ecology. 1999;**14**:493-507

[121] Drillet Z, Fung TK, Leong RAT, et al. Urban vegetation types are not perceived equally in providing ecosystem services and disservices. Sustainability. 2020;**12**:2076

[122] Magle SB, Hunt VM, Vernon M, et al. Urban wildlife research: Past, present, and future. Biological Conservation. 2012;**155**:23-32

[123] Collins MK, Magle SB, Gallo T, et al. Global trends in urban wildlife ecology and conservation. Biological Conservation; **261**. Epub ahead of print 2021. DOI:10.3389/fvets.2018.00014

[124] Flies EJ, JNW L, Douglas I. Editorial: Urban ecology and human health. Frontiers in Ecology and Evolution. 2022;**10**:1-3. Epub ahead of print 2022. DOI:10.3389/fevo.2022.1032022

[125] Ferrini F, Fini A, Mori J, et al. Role of vegetation as a mitigating factor in the urban context. Sustainability. 2020;**12**:4247

[126] Destoumieux-Garzón D, Mavingui P, Boetsch G, et al. The one health concept: 10 years old and a long road ahead. Frontiers in Veterinary Science. 2018;**5**:14

[127] Perry GT, Gebresenbet F, DaPra M, et al. Why urban ecology matters in Ethiopia. Frontiers in Ecology and Evolution. **10**. Epub ahead of print 2022. DOI:10.3389/fevo.2022.843698

[128] MacArthur RH, MacArthur JW. On bird species diversity. Ecology. 1961;**42**:594-598

[129] Smith RM, Thompson K, Hodgson JG, et al. Urban domestic gardens (IX): Composition and richness of the vascular plant flora, and implications for native biodiversity. Biological Conservation. 2005;**129**:312-322

[130] Lerman SB, Warren PS. The conservation value of residential yards: Linking birds and people. Ecological Applications. 2011;**21**:1327-1339

[131] Rega-Brodsky CC, Nilon CH, Warren PS. Balancing urban biodiversity needs and resident preferences for

vacant lot management. Sustainability. 2018;**10**:1679

[132] Riley CB, Perry KI, Ard K, et al. Asset or liability? Ecological and sociological tradeoffs of urban spontaneous vegetation on vacant land in shrinking cities. Sustainability. 2018;**10**:2139

[133] Mitchell JC, Jung Brown RE, Bartholomew B. Urban Herpetology. Salt Lake City: Society for the Study of Amphibians and Reptiles; 2008

[134] Jellinek S, Driscoll DA, Kirkpatrick JB. Environmental and vegetation variables have a greater influence than habitat fragmentation in structuring lizard communities in remnant urban bushland. Austral Ecology. 2004;**29**:294-304

[135] Van Helden BE, Close PG, Stewart BA, et al. An underrated habitat: Residential gardens support similar mammal assemblages to urban remnant vegetation. Biological Conservation. 2020;**250**:108760

[136] Van Helden BE, Close PG, Steven R. Mammal conservation in a changing world: Can urban gardens play a role? Urban Ecosystem. 2020;**23**:555-567

[137] Ofori BY, Garshong RA, Gbogbo F, et al. Urban green area provides refuge for native small mammal biodiversity in a rapidly expanding city in Ghana. Environmental Monitoring and Assessment. 2018;**190**:1-11

[138] Zungu MM, Maseko MS, Kalle R, et al. Effects of landscape context on mammal richness in the urban forest mosaic of EThekwini Municipality, Durban, South Africa. Global Ecology and Conservation. 2020;**21**:878

[139] Zungu MM, Maseko MS, Kalle R, et al. Factors affecting the occupancy of forest mammals in an urban-forest mosaic in EThekwini Municipality, Durban, South Africa. Urban For Urban Green. 2020;**48**:126562

[140] Brum PHR, Gonçalves SRA, Strüssmann C, et al. A global assessment of research on urban ecology of reptiles: Patterns, gaps and future directions. Animal Conservation. Epub ahead of print 15 June 2022. DOI:10.1111/acv.12799

[141] Melliger RL, Braschler B, Rusterholz HP, et al. Diverse effects of degree of urbanisation and forest size on species richness and functional diversity of plants, and ground surface-active ants and spiders. PLoS One. 2018;**13**:199245

[142] Mills JG, Bissett A, Gellie NJ, et al. Revegetation of urban green space rewilds soil microbiotas with implications for human health and urban design. Restoration Ecology. 2020;**28**:322-334

[143] Francoeur XW, Dagenais D, Paquette A, et al. Complexifying the urban lawn improves heat mitigation and arthropod biodiversity. Urban Forestry & Urban Greening. 2021;**60**:127007

[144] Hall DM, Camilo GR, Tonietto RK, et al. The city as a refuge for insect pollinators. Conservation Biology. 2017;**31**:24-29

[145] Bennett AB, Lovell S. Landscape and local site variables differentially influence pollinators and pollination services in urban agricultural sites. PLoS One. 2019;**14**:212034

[146] Cruz JC, Ramos JA, Silva LP, et al. Seed dispersal networks in an urban novel ecosystem. European Journal of Forest Research. 2013;**132**:887-897

[147] de Freitas TC, Gomes GC, Ramos Molina A, et al. Artificial perches increase bird-mediated seed rain in

agricultural fallow area in southern Brazil. Web Ecology. 2022;**22**:59-74

[148] U.S. Census Bureau. Evaluation of anuran persistence in an urbanized drought-affected setting in the southern High Plains. Population and Housing Unit Estimates. 2020. Available from: https://www.census.gov/quickfacts/fact/note/US/PST045221

[149] United States Climate Data. Climate Lubbock—Texas and Weather averages Lubbock. 2022. Available from: https://www.usclimatedata.com/climate/lubbock/texas/united-states/ustx2745 [Accessed: 14 January 2023]

[150] Ramesh R. Evaluation of Anuran Persistence in an Urbanized Drought-Affected Setting in the Southern High Plains. Lubbock, Texas: Texas Tech University; 2011

[151] Freeman S. Relationships of Structural Characteristics and Homeowner Socioeconomics with Urban Vegetation and Bird Communities. Lubbock, Texas: Texas Tech University; 2018

[152] Carter CR. Mesopredator Occurrence and Relationships with Natural and Anthropogenic Features in an Urban Area on the Southern High Plains of Texas. Lubbock, Texas: Texas Tech University; 2019

[153] Coman IA, Cooper-norris CE, Longing S, et al. It is a wild world in the City: Urban wildlife conservation and communication in the age of COVID-19. Diversity. 2022;**14**:539

[154] Entman R. Projections of Power: Framing News, Public Opinion, and US Foreign Policy. Chicago: University of Chicago Press; 2004

[155] Lecheler S, De Vreese CH. News Framing Effects: Theory and Practice. London and New York: Taylor & Francis; 2019

[156] Huang JH, Floyd MF, Tateosian LG, et al. Exploring public values through Twitter data associated with urban parks pre-and post-COVID-19. Landscape and Urban Planning. 2022;**227**:104517

[157] Tateosian LG, Saffer A, Walden-Schreiner C, et al. Plant pest invasions, as seen through news and social media. Computers, Environment and Urban Systems. 2023;**100**:101922

[158] Gianotti AGS, Hurley PT. Gathering plants and fungi along the urban-rural gradient: Uncovering differences in the attitudes and practices among urban, suburban, and rural landowners. Land Use Policy. 2016;**57**:555-563

Chapter 7

Perspective Chapter: Mechanization in Agricultural Production from Horizontal and Vertical Perspective

Mohd. Muzamil, Sehreen Rasool, Mohd. Mudasir Magray, Ummyiah H. Masoodi, Shabir Ahmed Bangroo, Ajaz Ahmed Malik and Saba Banday

Abstract

The mechanization of agricultural system, both horizontal and vertical cultivation, is imperative for judicious application of resources, reduction in drudgery of workforce, amelioration of productivity and improvement in competitiveness of the produce. However, the reduction in per capita land availability has triggered a mass migration towards vertical cultivation system with heavy reliance towards automation, Internet of Things (IoT) and artificial intelligence. The vertical system of cultivation and protected cultivation system is essential to overcome the limitations of small land holdings, particularly in developing countries and combat global climate change. With the result, the concept of hydroponics, aquaponics, aeroponics is gaining momentum at a rapid pace. The horizontal and vertical system also demands the preparation of organic fertilizer through advanced machinery for bolstering the soil fertility and enhancement in productivity of agricultural crops.

Keywords: compaction, urban agriculture, hydroponics, waste, greenhouse

1. Introduction

Agricultural system is believed to have touched 'tipping point'. The tipping point theory was promulgated as the system is witnessing significant and catastrophic transformation [1] due to the limitations induced by emergence of resistant pests (insects), extreme climatic events, unsustainable land management practices, poor soil conservation measures, attenuation of soil nutrients and global climate change. Many intellectuals believe that the expansion of the agricultural system in the last decade can be attributed to an increase in the production of major crops and cropping intensity as opposed to the expansion in landmass. The contribution of the expansion in the land mass can be as low as 10% [2].

Figure 1.
Mechanization potential in agricultural production system.

The production of the agricultural crop demands the involvement of mechanical interfaces in the completion of the unit agricultural activities. The mechanization is preferred owing to its involvement in the precision agriculture, automation, post-harvest operations, waste management and organic agriculture, **Figure 1**. The main objective of farm mechanization is to ensure judicious application of scarce resources, timeliness of the operations, reduction in the drudgery and lowering of input cost in the production system. The reduction in the production cost, directly or indirectly, favours the competitiveness of the produce at local, national and international markets. The production system involves several operations that must be completed within stipulated time to thwart losses and protect the crop from the vagaries of climate.

2. Mechanization in horizontal and vertical cultivation system

The mechanization of the agricultural operations is different for horizontal and vertical pattern of cultivation system. In horizontal system, the mechanical interfaces are more or less conventional, operated by prime mover (tractor, power tiller, electric motor, battery) on the levelled ground. In vertical system, most of the systems are protected within the boundaries or the ambient conditions are controlled to simulate the natural conditions. In such a case, the mechanization system is entirely different and demands a systematic approach to balance the mechanization of the horizontal pattern of cultivation.

2.1 Horizontal cultivation system

It is the process of cultivating a crop on a piece of land with the help of different unit operations and utilizing the same piece of land for another crop, when the former is harvested, **Figure 2**. The horizontal cultivation system is followed across the globe; however, the shrinkage of the land due to rapid urbanization has questioned the sustainability of the horticultural cultivation system. The decrease in per capita land holdings and quantum jump in world population has forced to search for an alternative model. The prediction of 8.3–10.9 billion global population by 2050 [3, 4] with 90% restricted within Asia and Africa has alarmed the scientific fraternity with respect to food security and environmental sustainability. Moreover, there is a steady decline in the productivity of the crops due to poor soil quality, low nutrient use efficiency, inappropriate water management strategies [5] and lack of arable land and freshwater [6].

2.1.1 Seedbed preparation

The seedbed preparation is cumbersome and consumes about 30% of the total energy in the production system. It involves subsoiling, primary and secondary tillage machinery. Subsoiling is the process of breaking the hard pan or plough sole formed

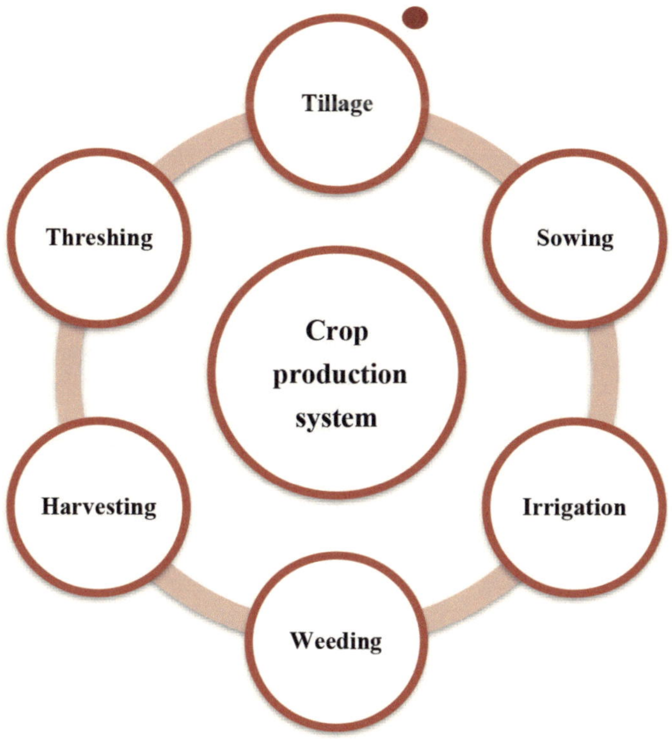

Figure 2.
Steps involved in the crop production system (first step).

after years of cultivation or movement of livestock or transportation machinery. The stress is usually transmitted by means of soil particles and converges at a single point. As the process continues, a layer of hard soil is formed that prevents the penetration of the roots and accumulates the water in the root zone, **Figure 3**. This process is known as soil compaction. It is recommended to use subsoiler after every 4–5 years to break the hard layer or plough sole. The subsoiler is a single-standard plough that penetrates deep within 45–100 cm or more and breaks the hard pan. It is operated by the >50 hp. tractor. However, when the hard pan or plough sole is formed within 20–25 cm depth, a chisel plough with many standards can be used. The subsoiling operation is followed primary tillage. The primary tillage is carried out with mould board plough and disc plough to reach to a depth of 20 cm. The movement of the tillage tool generates a triangular wedge that moves forward and causes breakdown of soil layers. The clods raised from the primary tillage undergo size reduction in secondary tillage operation in the range of 8–12 cm by cultivator, rotavator and disc harrow.

2.1.2 Sowing/transplanting

The process of placing the seeds or seedlings in the well-prepared seedbed is termed as sowing or transplanting. The metering system of the seeder contains a horizontal or vertical disc with grooves on the outer periphery to drop single seed at one time [7]. In case of transplanting operation, like in paddy or vegetables, a special type of mat seedling is used and finger-type transplanter picks one seedling at a time and plants it in the soil. A number of manual, power-operated and self-propelled seeders

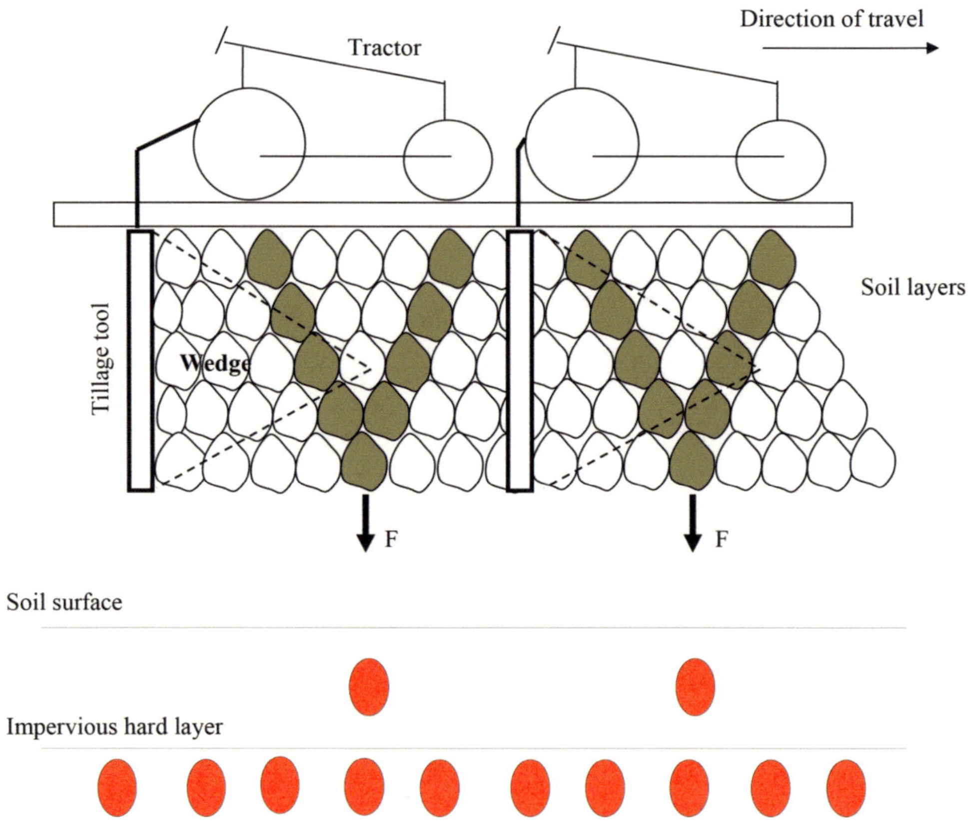

Figure 3.
Generation of stress concentrating points in the plough sole.

and transplanters are available for sowing or transplanting of seeds or seedlings at different rates and planting distances.

2.1.3 Irrigation

The seeds in the seedbed require sufficient moisture to sustain the metabolic activities and provide feasible conditions for the emergence of the plants. Traditionally, the irrigation is provided by flooding method; however, it results in wastage of water and possesses less efficiency. The modern system involves sprinkler and drip irrigation with sophisticated systems to ensure high water use efficiency and less wastage. The governmental agencies are also thrusting to involve sprinkler and drip irrigation system to ensure 'more crop per drop'. Recently, Internet of Things (IOT)-powered *in situ* real-time monitoring system has enabled to improve water use efficiency [8].

2.1.4 Weeding

The growth of the plants also results in the growth of unwanted weeds. These weeds interfere with the growing mechanism of the plants and create a competitive environment for the assimilation of nutrients, water and fertilizers. The weeds are removed by chemical methods and mechanical weeders. The mechanical weeders are manual, power-operated, self-propelled and battery operated, **Figure 4**.

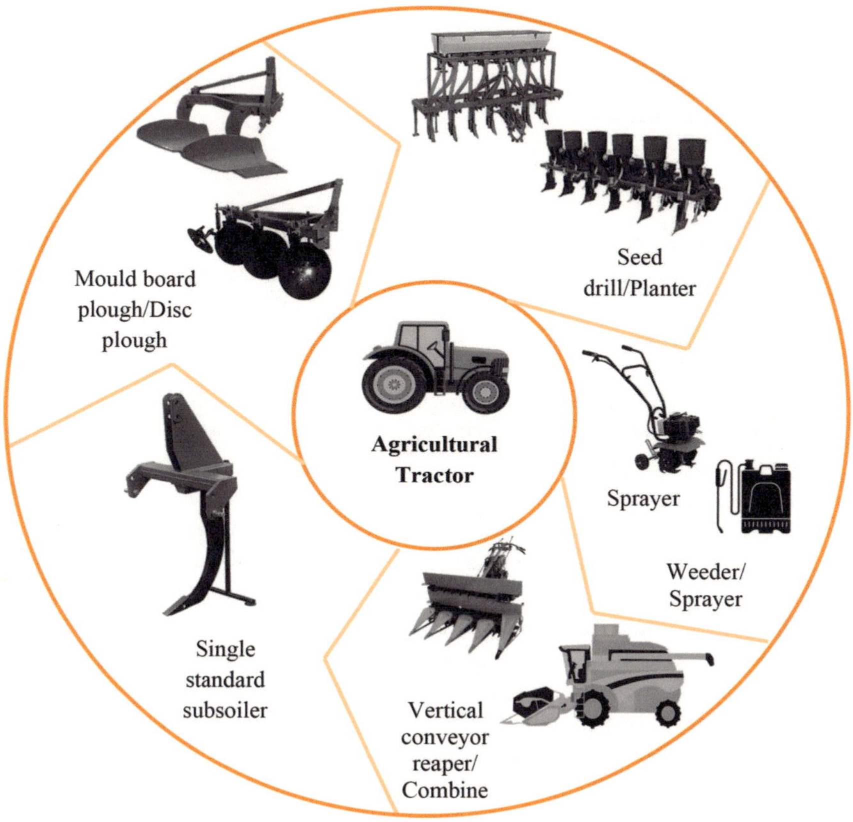

Figure 4.
Farm machines for horizontal cultivation system.

2.1.5 Harvesting

Harvesting is the process of reaping (cereal crops), plucking (fruit crops), digging (root crops) by manual or power-operated mechanical harvesters [9]. A power-operated mechanical harvester or combine comprised of a collection unit, threshing cylinders, cleaning mechanism, collecting system and chaff blowing unit. The crop dividers and reel intend to divert the standing crop towards the reciprocating cutter bar. The harvested crop is fed to the threshing cylinders through a conveying mechanism. The threshed grain passes through the reciprocating screen to separate the grains from the chaff. The clean grain is collected in the trays placed at the lower side. The chaff is blown and collected separately. The losses in combine are limited ranging from 2.5% in pulses and cereals and 4% for soybean.

2.1.6 Threshing

The harvesting operation is followed by threshing, which results in the segregation of plants and grains/fruits. However, these days, combine is often used as it can harvest, thresh, clean and collect in single operation. There are five types of threshers depending on the type of crop and usage: wire loop, spike tooth, syndicator type, hammer mill type and rasp bar type.

2.1.7 Post-harvest operations

The post-harvest operations intend to reduce the losses and enhance the shelf life of the agricultural produce.

2.1.8 Storage

The storage of the agricultural produce is the last step in the chronological process of production system to preserve and protect the crop for future use.

2.2 Vertical cultivation system

Vertical pattern of cultivation is perceived as the panacea of the ills of horizontal pattern of cultivation. The problems of small size, high labour requirement and demand for urban land have compelled to shift towards vertical farming system [10]. In this method, the space is utilized judiciously by creating stepwise layer system of cultivation. The vertical cultivation system is a vertically stacked layered system, where the grain crops, vegetables and fruit crops are grown in artificial or mechanically controlled conditions in cities and urban areas [11]. These vertical farms use a combination of solar panels to control the lighting system, temperature monitoring system, sensors, air humidity and maintenance unit to reduce the impact of environmental parameters and lower down the cost of production, **Figure 5**. It also uses soilless techniques to grow the plants by supplying the nutrient solution through the root zone of the plants [12]. The vertical cropping system results in an increase in the productivity of the crop per unit base area. According to Touliatos et al. [13], the production of lettuce was significantly (13.8 times) higher in vertical farming in comparison with horizontal cultivation system.

The benefits of the vertical farming system outnumber the benefits of horizontal farming system, **Table 1**. However, the main challenge is to reduce the cost and energy consumption, which is directly or indirectly responsible for increasing the carbon footprint. The cost can be splitted into capital cost used for fabrication of farm and operational cost for running the farm on daily basis. When the farm is small, major costs are incurred on capital costs; when the farm is large, it is tilted towards operating

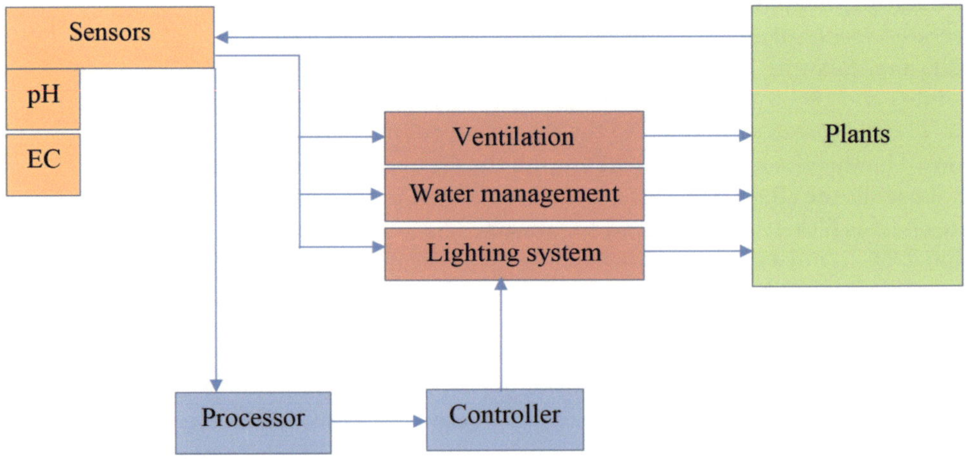

Figure 5.
Automated farm production system.

Type of system	Benefits	Reference
Vertical farming	Higher productivity	[13]
	Different crops in different layers	[6]
	Protection from vagaries of climate and natural disasters	[6]
	Multiple harvest round the year	[14]
	Reduction in the usage of fossil fuels	[6]
	Protection of environment from methane emissions	[15]

Table 1.
Benefits of vertical farming system.

costs *viz.* heating, ventilation and air conditioning (HVAC). Banerjee and Adenaeuer [16] modelled and concluded that the vegetable production in vertical farm requires 14 GWh of power per hectare per annum, which is much higher than 1.75 GWh per hectare per annum in horizontal farming [17]. Moreover, the initial fabrication installation cost of vertical farming system is higher than conventional agricultural setup.

3. Urban agriculture

Urban agriculture is often discussed owing to rising poverty, food shortages and nutritional deficiency in the urban cities. It can be inferred as mixed system of vegetables, fruits, trees and condiments with main thrust on reducing the expenditure and augmenting the income. The urban agricultural system has the proclivity to supply 8% proteins and 40% calcium intake to poor city dwellers [18]. It is a viable approach towards reducing the vulnerability, survivability and food security of the urban poor. Urban agriculture offers multiple benefits (**Figure 6**), particularly in terms of sustainability, accessibility to green spaces, recreation activities, health benefits, inclination towards nature, reduction in transportation cost, mitigation of adverse climatic events, promotion of resource use efficiency and income augmentation [19]. However, the city lands are often used for activities that produce the highest net income [20]. Accordingly, the preference for the allocation of the land is mostly non-agricultural activities. The diversion has the inclination to jeopardize food security and augment poverty levels [21]. The discrimination has pushed the agriculture towards the outer periphery of the urban areas, a term called peri-urban agriculture [22].

Peri-urban agriculture, implying agricultural setup in the midst of cities and rural areas, is characterized by poor infrastructure, less human population density and

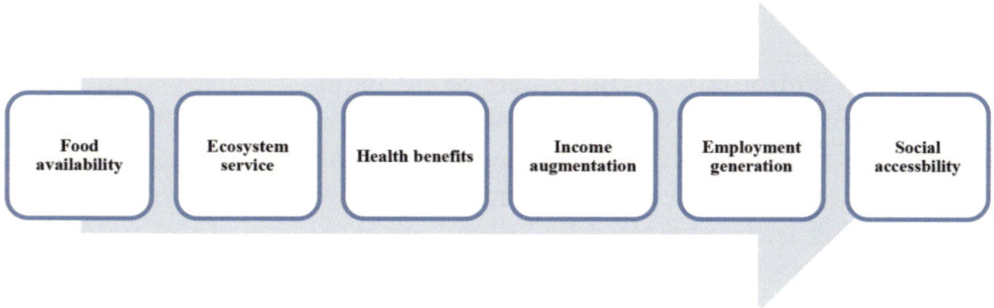

Figure 6.
Services offered by the urban agriculture.

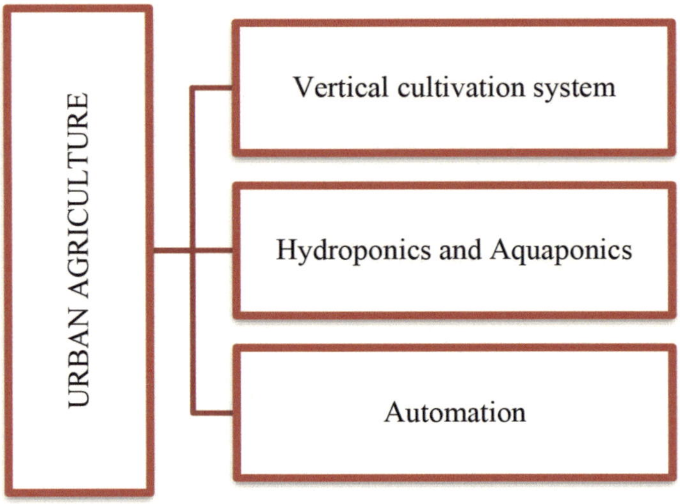

Figure 7.
Technologies used in urban agriculture.

availability of large areas for agricultural activities [23]. However, the main thrust of peri-urban agriculture is income generation as the associated farmers are mainly professionals [24]. The urban agriculture is categorized into three categories depending on the type of technology used for the cultivation of crops, **Figure 7**.

The proper composition of physiological and environmental parameters is indispensable for the growth and development of plants. These parameters include temperature, light, relative humidity and nutrient availability; presence of moisture dictates the quality of the plant and productivity of the crop. These parameters are controlled in hi-tech greenhouses and plant factories [25].

3.1 Hydroponics

The traditional horizontal agricultural system requires soil to grow the crops. In hydroponics, the crops are grown in natural or artificial substrates (soil free environment) and the nutrients (micro and macro) are supplied to the root zone of the plants [26]. The substrates can be organic, inorganic or synthetic [27] with different properties, **Table 2**. The commonly used fertilizers in hydroponic system include ammonium nitrate (NH_4NO_3), phosphoric acid (H_3PO_4), calcium nitrate {$(CaNO_3)_2$} and nitric acid (HNO_3). The chemical formulations are sold in three-digit sequence of nitrogen (N), phosphorus (P) and potassium (K) *viz.* 8–15–32 implying 8% N, 15% P and 32% K. It is also important to replenish the water in the nutrient solution continuously to keep the nutrients appropriate to the growth of the plants. In fact, the nutrient solution in the tank should be changed, and tank should be cleaned and disinfected after every 2 to 3 weeks [30]. The system also uses sensors, software, microcontrollers, mobile applications, web platforms and computing devices to control the climatic parameters, lighting and irrigation scheduling. The expansion of the hydroponic system is farfetched and the market is expected to grow at 20.7% CAGR from 2021 to 2028 [31]. Hydroponics is termed as the best alternative for urban and peri-urban agriculture and a futuristic approach to achieve 'sustainable cities and communities' goal of United Nations' Agenda for Sustainable Development 2030.

Substrate		Properties	Reference
Organic	• Peat • Coconut fibre • Sand • Pumice • Vermiculite	• pH	—
		• Electrical conductivity	—
		• Bulk density	[28]
		• Porosity	[28]
Inorganic	• Stonewool • Zeolites • Expanded clay • Perlite	• Water holding capacity	[29]
		• Cation exchange capacity	[28]
Synthetic	• Polystyrene • Polyurethane foam	• Cost	-

Table 2.
Substrates used in hydroponics.

The basic structure of the hydroponic system comprises of a perforated tray, reservoir, pump, delivery tubes, aerator/air pump and lighting system. The selection of the hydroponic systems depends on environmental conditions, cost, type of the crop, level of technology and acceptability, **Figure 8**. Hydroponic system results in higher yield and lower water consumption in lettuce [32] and higher plant survival in strawberries [26]. However, the energy consumption is much higher [32] and waste water treatment is minimal.

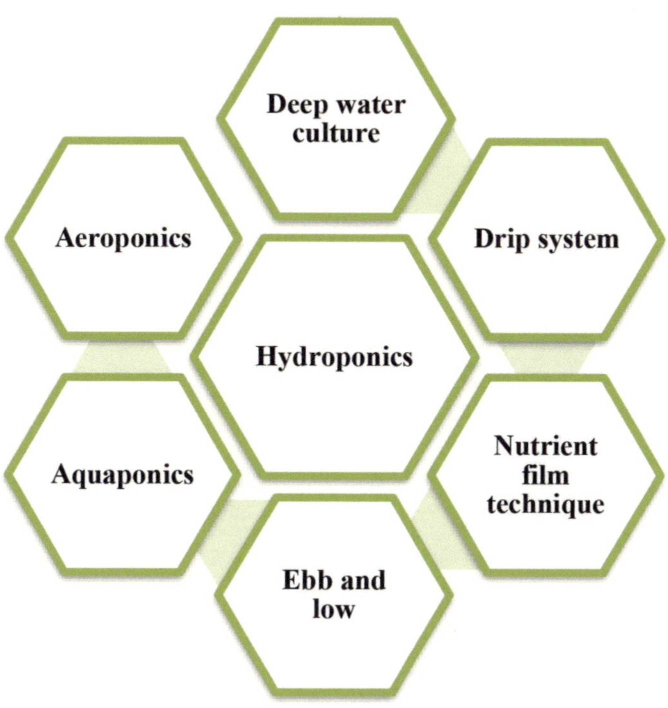

Figure 8.
Types of hydroponic system.

3.1.1 Deep water culture

The storage tank with nutrient solution contains spaces to place the plants and an aerator to circulate the oxygen continuously. The roots of the plants are immersed in the nutrient solution, while the upper half remains outside the tank with the help of wood, polystyrene or cork bark.

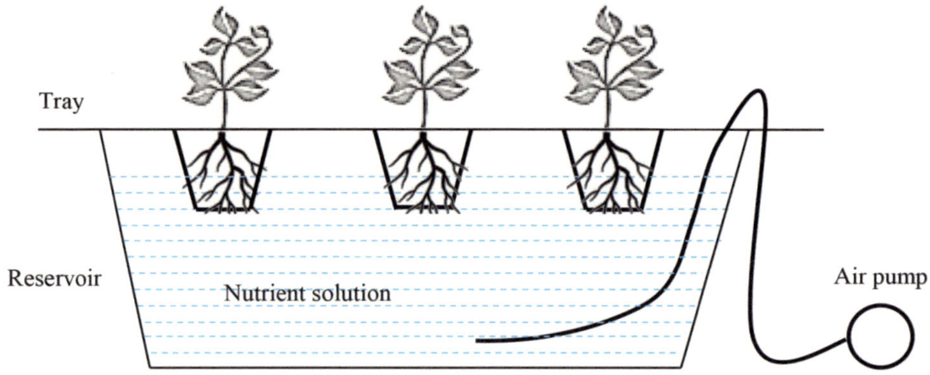

3.1.2 Drip system

It is similar to deep water culture; however, the nutrient solution is pumped through a common channel that contains the roots of the plants. The solution is supplied at specific time intervals in controlled flow and the residual solution is bypassed to the storage tank/reservoir. This method is suitable for specific crops *viz.* tomato and pepper.

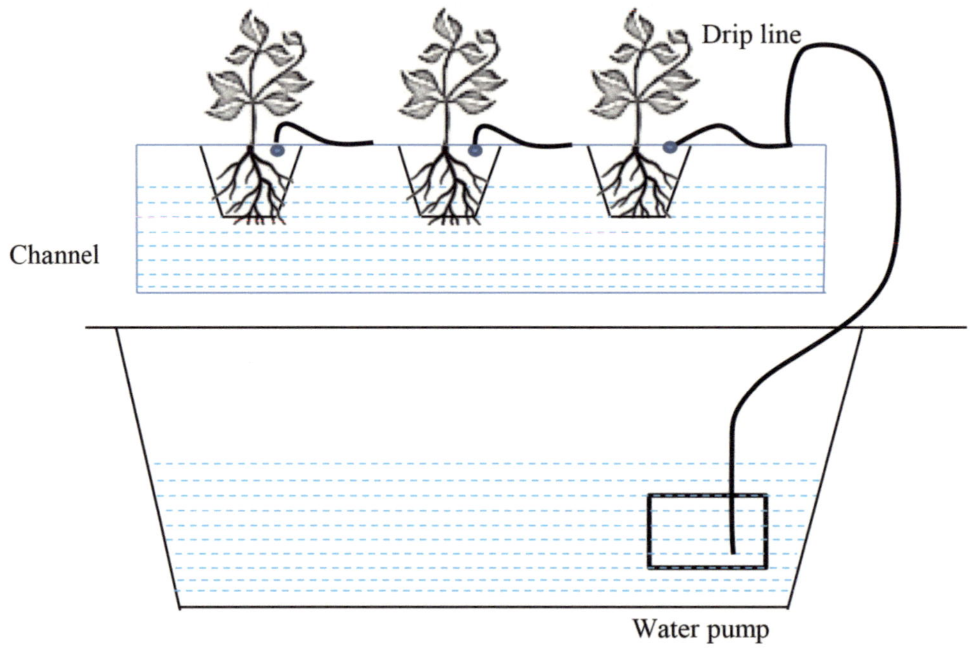

3.1.3 Nutrient film technique

This technique relies on supplying the nutrient solution to the roots of the plant continuously. A number of trays can be placed one above the other to reduce the space constraints and increase the productivity of crops.

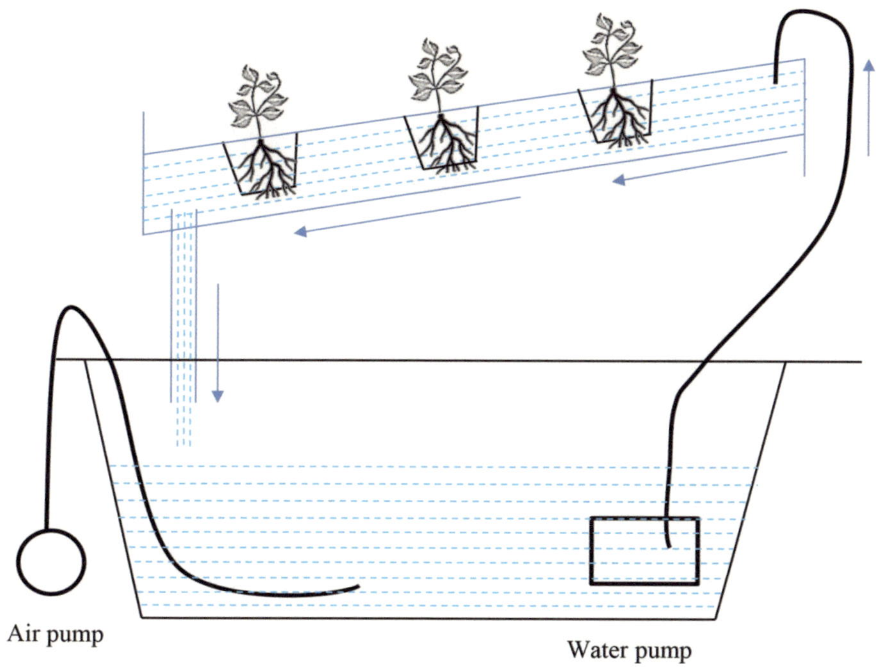

3.1.4 Ebb and low

It is a system in which the plants are kept in a tray filled with nutrient solution. The nutrient solution is pumped from the tank placed below the tray. The recycling of water is ensured by the force of gravity.

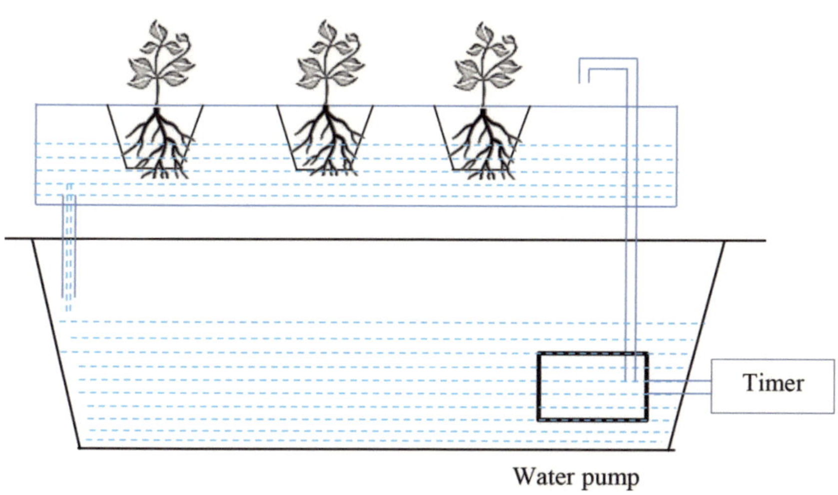

3.2 Aquaponics

The aquaponics establishes symbiotic relationship between fish and plant growth. The sludge of the fish, rich in nutrients, is used as the nutrient medium to support the growth of the plants. This forms a sustainable ecosystem in which the plants grow on one side and microbial process of nitrification/denitrification on the other side.

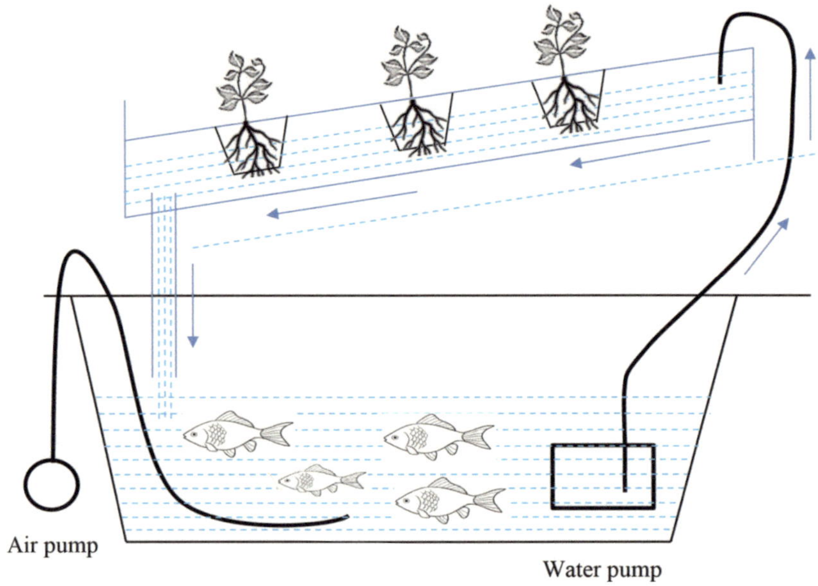

3.3 Aeroponics

'Aero' means air *i.e.* the nutrient solution is absorbed by the plant roots in air. In this method, the nutrient solution is sprayed by means of sprinklers in a common zone and the roots absorb the nutrients from the air. It is advantageous as the roots get aerated from the oxygen present in the sprayed solution.

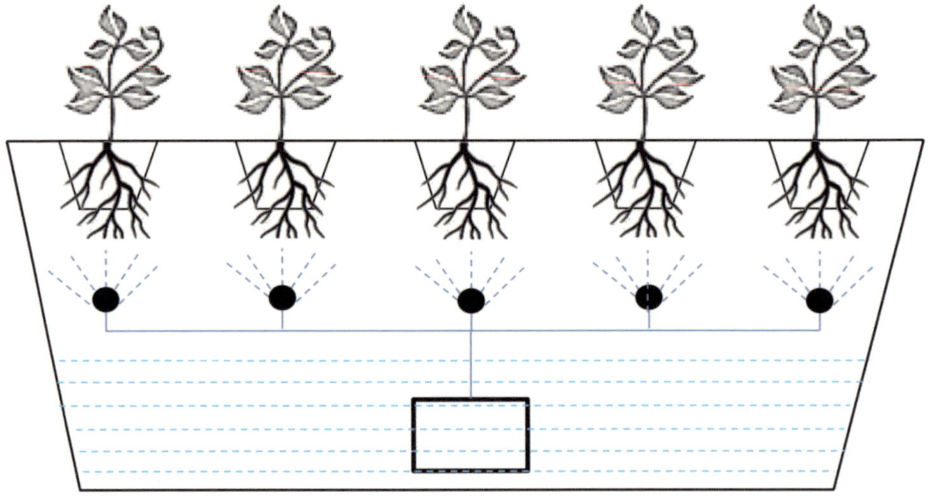

However, there are certain challenges that must be rectified to ensure the sustainability and viability of hydroponic system. The initial cost of fabrication, installation and equipment is high. The operator must be skilled with sufficient knowledge of agriculture, electronics, plant physiology and plant pathology. Moreover, the residual nutrient solution must be properly disposed as it may pollute the ecosystem.

3.4 Automation in vertical cultivation

Modern technological techniques such as artificial intelligence and machine learning-based neural network, deep learning, fuzzy logic, big data and Bayesian network are employed to improve the working performance of the hydroponic system [33]. Internet of Things (IOT) is believed to have a number of applications in vertical cultivation system. It integrates different sensors with microcontrollers and Wi-Fi modules to manage the resources in real time [8]. It helps in reducing the manual errors and improves the productivity of the crops. The biosensors can provide real-time data that help in detection of diseases and pests [34] and make logical decisions [35].

4. Protected cultivation system

The greenhouses are used for raising of crops, mainly to protect them from harsh climatic conditions and extension of the cultivation season, **Figure 9**. It integrates different renewable sources to promote and sustain crop growth [36]. Solar energy for electricity [36] and drip irrigation for water conservation [37, 38] improves the economic viability of the greenhouse cultivation system [39]. In fact, greenhouse irrigation sensor-based automation system is perceived as the effective method to prevent the wastage of water [40].

It also provides a window of opportunity to cultivate off-season crops. The greenhouses are usually made from plastic, glass or polyethylene to allow the radiations

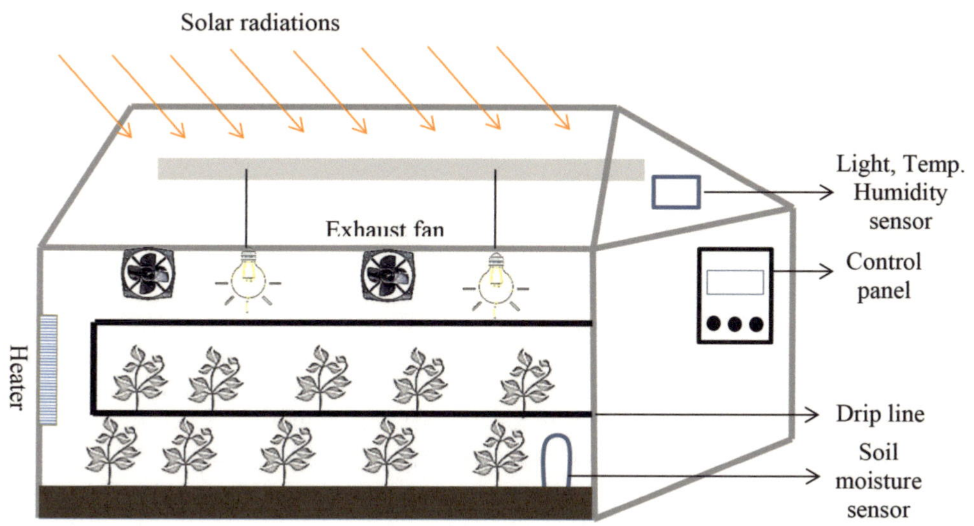

Figure 9.
Automated greenhouse structure.

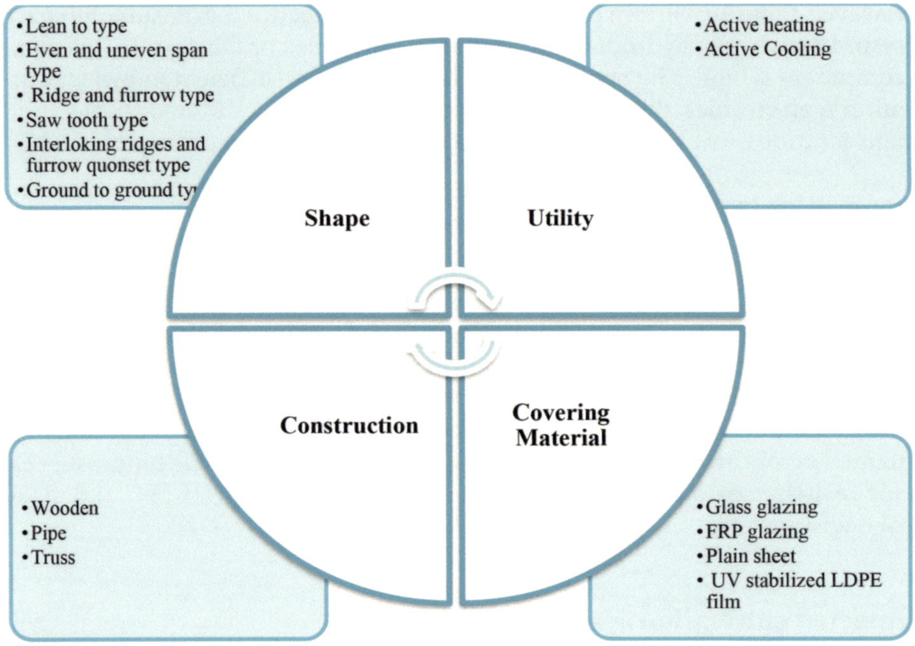

Figure 10.
Types of greenhouses.

to pass through and contribute towards plant growth. The greenhouse can be open, semi-closed and closed type. In open type, there is no provision to collect and reuse the drained water. The semi-closed type has small windows with low cooling capacity and mechanical ventilation system. The semi-closed type reuses the drained nutrient solution by collecting it in a reservoir that is topped-up with fresh water on regular basis. In closed type, the water follows a closed loop for collection, distribution, recycling and re-distribution. The mechanical system is used for cooling and dehumidification by air treatment accessories.

The polyhouse's are categorized (**Figure 10**) on the basis of shape, utility, construction and covering material [41]. The cost of the construction is an essential parameter of the poly house. Accordingly, it is categorized into low-cost green house, medium-cost green house and high-cost green house.

5. Waste management

Urban agriculture relies on utilizing plant residues and kitchen waste for production of bio-fertilizer [42]. The wastes can be converted into compost or vermicompost for round-the-year agricultural production activities. However, the space is the limiting factor in the urban areas [43]. The involvement of low-cost smart vermicomposting bin with automatic watering and mixing system can help to transform the wastes into fortified vermicompost [44]. A single large space can also be devoted towards the preparation of compost through windrow composting [45]. In windrow composting, the degradable materials are laid in the form of long windrows, microbial culture is added to increase degradation rate and mixed at regular intervals (**Figure 11**) with the help of windrow turner [45].

Figure 11.
A tractor-operated windrow turner in working mode [45].

6. Conclusion

The shrinkage of the land and increased wages has compelled to shift towards mechanization in both horizontal and vertical cultivation systems. The mechanical systems are undergoing a transformation towards automation to reduce the manual errors and improve their efficiencies. On the other hand, the vertical mechanical systems and automations might witness a reduction in the installation, operational and maintenance cost for higher adoptability among the stakeholders. The success of the mechanical farming systems might be directly linked with the security, sustainability and environmental viability of the world in near future.

Acknowledgements

The support and help rendered by the officials at SKUAST-K and IARI is highly appreciated and duly acknowledged.

Conflict of interest

The manuscript was prepared unanimously. Hence, no conflict of interest exists.

Author details

Mohd. Muzamil[1*], Sehreen Rasool[2], Mohd. Mudasir Magray[3], Ummyiah H. Masoodi[4], Shabir Ahmed Bangroo[5], Ajaz Ahmed Malik[5] and Saba Banday[6]

1 College of Agricultural Engineering and Technology, SKUAST-K, India

2 College of Temperate Sericulture, Mirgund, SKUAST-K, India

3 Division of Horticulture, Faculty of Agriculture, SKUAST-K, India

4 Division of Vegetable Science, Faculty of Horticulture, SKUAST-K, India

5 Division of Soil Science, Faculty of Horticulture, SKUAST-K, India

6 Division of Plant Pathology, Faculty of Horticulture, SKUAST-K, India

*Address all correspondence to: muzamil4951@gmail.com

References

[1] Barnosky AD, Hadly EA, Bascompte J, Berlow EL, Brown JH, Fortelius M, et al. Approaching a state shift in Earth's biosphere. Nature. 2012;**486**:52-58

[2] Alexandaratos N, Bruinsma J. World Agriculture Towards 2030/50: The 2012 Revision, ESA Work Paper 12-03. Rome: UN Food Agriculture Organization (FAO); 2012

[3] Bringezu S, Schutz H, Pengue W, Obrein M, Garcia F, Sims R. Assessing Global Land Use: Balancing Consumption with Sustainable Supply. Nairobi/Paris: UNET/International Resource Panel; 2014

[4] Muzamil M, Rasool S, Masoodi UH. Insitu and exsitu agricultural waste management system. Agricultural Waste-New insights. 2022. DOI: 10.5772/intechopen.108239

[5] Dawe D, Dobermann A, Moya P, Abdulrachman S, Singh B, Lal P, et al. How widespread are yield declines in long term Rice experiments in Asia. Field Crops Research. 2000;**66**(2):175-193. DOI: 10.1016/S0378-4290(00)00075-7

[6] Mok WK, Tan YX, Chen WN. Technology innovations for food security in Singapore: A case study of future food systems for an increasingly natural resource-scarce world. Trends in Food Science and Technology. 2020;**102**:155-168

[7] Malla UM, Khan AR, Muzamil M, Mohiuddin M, Rashid S, Banday RZ, et al. Development and evaluation of four-row seeder for vegetable crops in the temperate region of Jammu and Kashmir. SKUAST Journal of Research. 2022;**24**(1):71-79. DOI: 10.5958/2349-297X.2022.00016.2

[8] Banday R, Muzamil M, Kumar A, Mohiuddin M, Kumar R, Rashid S. Internet of things based smart farming: A futuristic approach. Agricultural Engineering Today. 2023;**46**(4):41-44

[9] Muzamil M, Mani I, Kumar A, Shukla L. An engineering intervention to prevent paddy straw burning through insitu microbial degradation. Journal of Institution of Engineers. 2021;**102**:11-17

[10] Appiah DO, Asante F, Nketia BA. Perceived agricultural land use decisions in a peri-urban district, Ghana. Journal of Agricultural and Crop Research. 2017;**5**:1-10

[11] Al-Chalabi M. Vertical farming: Skyscraper sustainability? Sustainable Cities and Society. 2015;**18**:74-77. DOI: 10.1016/j.scs.2015.06.003

[12] Benke K, Tomkins B. Future food-production systems: Vertical farming and controlled-environment agriculture. Sustainability: Science, Practice and Policy. 2017;**13**(1):13-26. DOI: 10.1080/15487733.2017.1394054

[13] Touliatos D, Dodd IC, McAinsh M. Vertical farming increases lettuce yield per unit area compared to conventional horizontal hydroponics. Food and Energy Security. 2016;**5**(3):184-191. DOI: 10.1002/fes3.83

[14] Germer J, Sauerborn J, Asch F, Boer J, Schreiber J, Weber G. Sky farming an ecological innovation to enhance global food security. Journal für Verbraucherschutz und Lebensmittelsicherheit. 2011;**6**:237-251. DOI: 10.1007/s00003-011-0691-6

[15] Williams H, Wikström F. Environmental impact of packaging and

food losses in a life cycle perspective: A comparative analysis of five food items. Journal of Cleaner Production. 2011;**19**(1):43-48. DOI: 10.1016/j.jclepro.2010.08.008

[16] Banerjee C, Adenaeuer L. Up, up and away! The Economics of Vertical Farming. 2014;**2**(1):21. DOI: 10.5296/jas.v2i1.4526

[17] Himanshu S, Kumar S, Kumar A, Gupta KK. Energy economics assessment of crops in traditional and mechanized farming. International Research Journal of Environment Sciences. 2012;**1**(5):2319-1414

[18] Prain G, Lee-Smith D. Urban Agriculture in Africa: What Has Been Learned? In: Prain G, Lee-Smith D, Karanja N, editors. African Urban Harvest. New York: Springer; 2010. DOI: 10.1007/978-1-4419-6250-8

[19] Turner B. Embodied connections: Sustainability, food systems and community gardens. Local Environment. 2011;**16**:509-522

[20] Fisher EM, Fisher RM. Urban Real Estate. New York, NY, USA: Holt; 1954

[21] Hoornweg, D. and Munro-Faure, P. 2012. Urban Agriculture for Sustainable Poverty Alleviation and Food Security.

[22] Azunre GA, Amponsah O, Peprah C, Takyi SA, Braimah I. A review of the role of urban agriculture in the sustainable city discourse. Cities. 2019;**93**:104-119. DOI: 10.1016/j.cities.2019.04.006

[23] Opitz I, Berges R, Piorr A, Krikser T. Contributing to food security in urban areas: Differences between urban agriculture and peri-urban agriculture in the global north. Agriculture and Human Values. 2016;**33**:341-358. DOI: 10.1007/s10460-015-9610-2

[24] Armstrong D. A survey of community gardens in upstate New York: Implications for health promotion and community development. Health & Place. 2000;**6**(4):319-327

[25] Ting KC, Lin T, Davidson PC. Integrated urban controlled environment agriculture systems. In: Kozai T, Fujiwara K, Runkle ES, editors. LED Lighting for Urban Agriculture. Singapore: Springer Singapore; 2016. pp. 19-36

[26] Treftz C, Omaye ST. Comparison between hydroponic and soil systems for growing strawberries in a greenhouse. International Journal of Agricultural Extension. 2016;**3**(3):195-200

[27] Maucieri C, Nicoletto C, Os EV, Anseeuw D, Havermaet RV, Junge R. Hydroponic technologies. In: Goddek S, Joyce A, Kotzen B, Burnell GM, editors. Aquaponics Food Production Systems. Cham: Springer; 2019. DOI: 10.1007/978-3-030-15943-6_4

[28] Wallach J. Physical characteristics of soilless media. In: Raviv, Lieth, editors. Soilless Culture, Theory and Practice. Amsterdam: Elsevier; 2008. pp. 41-116

[29] Kipp JA, Wever G, DeKreij C. International Substrate Manual. Amsterdam: Elsevier; 2001

[30] Bosques JH. Curso Basico de Hidroponia. CA, USA: lulu.com Edit; 2010

[31] GVR. 2021. Hydroponics market size (2021-2028). Available from: https://www.grandviewresearch.com/industry-analysis/hydroponics-market/methodology.

[32] Barbosa GL, Gadelha FDA, Kublik N, Proctor A, Reichelm L, Weissinger E, et al. Comparison of land, water, and energy requirements of lettuce grown using hydroponic vs conventional

agricultural methods. International Journal of Environmental Research and Public Health. 2015;**12**(6):6879-6891. DOI: 10.3390/ijerph120606879

[33] Chowdhury MEH, Khandakar A, Ahmed S, Al-Khuzaei F, Hamdalla J, Haque F, et al. Design, construction and testing of a IoT based automated indoor vertical hydroponics farming test bed in Qatar. Sensors. 2020;**20**:5637

[34] Teh, C. 2019. Tiny Sensors Allowing Observation of Plant Growth May Help Urban Farms. The Straits Times. 17/08/2019. Available from: https://www.straitstimes.com/singapore/tiny-sensors-allowing-observation-of-plant-growth-may-help-urban-farms

[35] Ray PP. Internet of things for smart agriculture: Technologies, practices and future direction. Journal of Ambient Intelligence and Smart Environments. 2017;**9**:395-420. DOI: 10.3233/AIS-170440

[36] Karanisa T, Achour Y, Ouammi A, Sayadi S. Smart greenhouses as the path towards precision agriculture in the food-energy and water nexus: Case study of Qatar. Environment Systems and Decisions. 2022;**42**:521-546

[37] Hong Jun L, Ming Z, Zhen HZ, Xin L, Jian XX, Hong WP. Growth and quality of greenhouse tomato under cycle aerated subsurface drip irrigation. Journal of Drainage and Irrigation Machinery Engineering. 2015;**33**(3):253-259

[38] He J, Zhou L, Yao Q, Liu B, Xu H, Huang J. Greenhouse and field-based studies on the distribution of dimethoate in cotton and its effect on *Tetranychus urticae* by drip irrigation: Dissipation and control efficacy of dimethoate on *T. urticae* by drip irrigation. Pest Management Science. 2018;**74**(1):225-233. DOI: 10.1002/ps.4704

[39] Warner LA, Lamm AJ, White SA, Fisher PR, Beattie PN. A new perspective on adoption: Delivering water conservation extension programming to nursery and greenhouse growers. Journal of Agricultural Education. 2020;**61**(1):172-189

[40] Cakir R, Kanburoglu-Cebi U, Altintas S, Ozdemir A. Irrigation scheduling and water use efficiency of cucumber grown as a spring-summer cycle crop in solar greenhouse. Agricultural Water Management. 2017;**180**:78-87. DOI: 10.1016/j.agwat.2016.10.023

[41] Shukla A, Sharma A, Kant K. Solar greenhouse with thermal energy storage: A review. Current Sustainable/Renewable Energy Reports. 2016;**3**:58-66

[42] Mees C, Stone E. Food, homes and gardens: Public community gardens potential for contributing to a more sustainable city. In: Viljoen A, Wiskerke JSC, editors. Sustainable Food Planning. Wageningen: Ed. Wageningen Academic; 2012. pp. 431-452

[43] Trussell B. The Bid Rent Gradient Theory in Eugene, Oregon: An Empirical Investigation. Oregon: Department of Economics, University of Oregon; 2010

[44] Kumar A, Muzamil M, Dixit J. Smart vermicomposting bin for rapid transformation of dal Lake aquatic weed into fortified vermicompost. Journal of Recycling of Organic Waste in Agriculture. 2022;**12**:221-233. DOI: 10.30486/ijrowa.2022.1955835.1445

[45] Muzamil M, Mani I, De D, Lande S. Influence of operational parameters of windrow turner for mass production of compost. International Journal of Agricultural and Statistical Sciences. 2013;**9**:9-28

Chapter 8

Landscaping Promoting Sustainable Comfort in Cities

Fernanda Da Conceição Moreira and Eglerson Duarte

Abstract

The landscaping in cities provides, continuously, people with high visual, emotional and physical performance with access to different plant morphological architectures, leaves, trunks, flowers, fruits and seed colors. The shadows caused by the leaves of the trees allows a comfortable atmosphere with the temperature reduction, air humidity increase and environment aromatization of gardens, parks and streets by the volatile organic compound emission by leaves, stems and flowers. The combinations of these environmental factors improve the plant, animal and Homo sapiens development. The plant architectural distribution in high-slope areas minimizes problems such as flood sand landslides in urban centers, being an economically viable and sustainable alternative for heritage preservation, the physical structure soil and optimization of water flow. The landscaping of cities is the perfect setting for meeting people, recording videos and photos for personal archives, social media, developing personal and business marketing projects and network.

Keywords: garden, healthy life, urban projects, architecture, nature preservation

1. Introduction

The sustainable landscaping is the balance between the beauty of urban space and the preservation of nature. Projects with sustainability use techniques to build the beauty of the scenery efficiently using natural and artificial resources with practices that minimize the use of substances harmful to health and the environment [1].

The sustainable gardens provide a comfortable environment with a pleasant climate, thermal stability, soil covered with plants and high air quality. Sustainable landscaping makes it possible to increase the longevity of people's lives as it allows interaction with a healthy natural environment, ideal for leisure, sports, relaxation and stress relief [2]. One of the goals of sustainability in landscaping is to contribute to people's awareness of the importance of preserving nature [3].

Urbanization is the development of cities, the construction of houses, buildings, condominiums, hospitals, schools, universities, parks, streets, avenues and viaducts, basic sanitation and electricity networks. This development occurs with population growth in a given location, improving the structure of small towns and large urban centers. In planning the landscaping project, solutions must be sought for social, economic and nature destruction problems in large urban centers and also in small towns where there is no awareness of the importance of preserving nature [3, 4].

IntechOpen

The sustainable landscaping has, socially, the green area as a place of education for people by creating spaces integrating buildings and plants. Culturally, sustainable landscaping projects aim to preserve heritage by expressing the local natural memory, harmoniously integrating the green area with the original characteristics of existing buildings, native vegetation, mineral resources and works of art [5].

Economically, sustainable landscaping selects materials with a high quality standard certified by international quality standards for the construction of concrete structures, lighting, water drainage networks and decoration. The selected plants can be native and plants adapted to the climate of the place where the sustainable landscaping project will be built and cultivated [3].

Developing sustainable landscaping projects, planting and building gardens, educating and encouraging the use of plants in homes, streets and parks is a noble mission and helps to keep nature alive, green and blooming in cities. Small gardens can be built in public spaces, as many cities have urban voids that could be used as spaces for living together and contemplating the fauna and flora of the region [6, 7].

In large urban centers, the huge surface of concrete and asphalt in the hot season release heat and increase the temperature. To reduce the effects of this event, ecological roofs known as the green roof are built, which is a covering of plants in buildings with waterproofing and cultivation techniques, being built by a team of qualified professionals. Green roofs have gained more and more space in urbanization projects in large cities, where the square meter of land generally has a high financial value [2].

On the green roofs, several species of plants are cultivated with a high capacity for shading and cooling the atmosphere, reducing the temperature and increasing the relative humidity of the air, promoting thermal comfort and beauty to be contemplated, being seen from the air, by pedestrians, cyclists and passengers by car and bus [8].

In buildings with a green roof, the internal temperature of the apartments and collective spaces is pleasant and there is energy savings with less use of air conditioning, increasing sustainability in the landscaping. Plants provide oxygen, make the environment more humid and attenuate noise, reducing the need for investment in acoustic insulation [4].

Vertical gardens optimize space when there is a reduction in the horizontal surface available for garden construction. In vertical garden projects, vases, green roofs must be positioned and cultivated in spaces with incidence of light, radiation, ventilation and water drainage system for irrigation of plants. Shady spaces can be ornamented with plants that are more efficient in using light [2].

For the garden to remain beautiful and healthy, periodic maintenance is necessary using techniques of fertilization, irrigation, pest control, pruning and removal of invasive plants. Sustainable practices reduce the use of irrigation, pesticides, fertilizers and conserve the soil by preventing erosion and sedimentation. In landscaping projects with artificial lakes, water filters must be constantly maintained to ensure water quality [3].

The heating of the real estate market in recent years and the growing desire of people to live surrounded by quality properties designed to meet expectations and environmental requirements. Landscapers, real estate developers, construction companies and public bodies have preferred the use of large native trees due to the rapid appreciation that these compositions bring to the urban environment and because they are efficient in restoring the balance of the local ecosystem [9]. The offer of sustainable gardening and landscaping services has increased and companies have specialized in sustainability to offer customers innovative and sophisticated landscaping projects, always aiming to promote people's well-being [7].

2. Tree shadows

Trees provide shade to the environment, beautify, refresh, offer shelter to animals (**Figure 1**). In urban landscaping to plant trees there is the process of planning space, tree species, maintenance, tree height and root depth [6].

The trees have several important functions such as producing, such as fruits, flowers, shade and good thermal sensation under their large branches and leaves (**Figure 2**). Fast growing shade trees are preferred in urban landscaping. There are several species of trees that grow quickly and provide excellent shade, such as the *Brunfelsia uniflora*, *Tibouchina granulosa*, and *Commiphora myrrha* [3].

Figure 1.
Trees in sustainable landscaping and walkways.

Figure 2.
Architecture of squares with trees for people to live together.

2.1 Comfortable atmosphere

In landscaping, flowers enchant us mainly for their vibrant colors and natural scent (**Figure 3**). The flowers of the plants exude a pleasant scent. The scent of the flowers serves to attract the sense of smell and bring the attractant closer to the flowers. The evolutionary target of these attractive resources to people and the perfume and cosmetics industry is not the human being. Bees, butterflies and flies are the target that flowers want to reach with the scent for noble reasons. These animals are the main pollinators of plants. When they land from flower to flower, they spread the pollen and ensure the fertilization of flowers in other places, increasing the genetic variability of the plants. There are also other forms of pollination, such as wind and water, but animals are the most important agents. They also benefit from this contact with flowers, since they can feed on their nectar [3, 10].

Bees are attracted to blue and yellow flowers. This is due to the fact that these animals see ultraviolet light. In addition, they prefer flowers that smell like fermenting food [1].

Figure 3.
Garden with vibrantly colored flowers.

Figure 4.
Butterfly pollinating the flower.

The showy flowers usually attract birds, as they have great eyesight. The odor of these plants, however, is not strong, since birds do not have a keen sense of smell. Another characteristic of the flowers visited by birds is the large amount of nectar found in them, thus ensuring adequate nutrition for the animal [3].

Flowers pollinated by butterflies are also showy, usually red, yellow and blue (**Figure 4**). In addition, these flowers make their nectar accessible (**Figures 5** and **6**) only to some species that have a long mouthparts [3]. Flowers with the smell of ripe fruit and little coloration are normally pollinated by bats. As these animals have nocturnal habits, the color of the flowers would not be a form of attraction. Flowers pollinated by beetles are also less showy and usually white [3].

Plants are sessile organisms and to compensate for their immobility, they have developed several mechanisms of interaction with the environment, including the release of volatile organic compounds into the soil and atmosphere from roots, fruits, flowers and leaves. The carbon fixed by plants is emitted back into the atmosphere

Figure 5.
Bees pollinating the flowers.

Figure 6.
Bird visiting the flowers.

in the form of volatile organic compounds. These emissions by vegetation play a fundamental role in the chemistry of the atmosphere, mainly in the aroma of the atmosphere [11].

The effect of absorbing the aroma in the human organism is fast, causing the body to respond both physically and in the deepest energies and emotions. The sense of smell is the most refined of the senses, guaranteeing the profound effect of aromatherapy, because when you feel a specific smell, it immediately awakens your emotions. Therefore, aromas are able to balance health, increase the feeling of happiness and awareness. Flowers have a higher vibration than we do, as they have no mental capacity. When a person comes into contact with a flower or a plant, their energy vibration also rises, favoring the maintenance and restoration of health [1].

3. Enhancement of comfort for high performance life development with sustainable landscaping

The combination of environmental factors optimizes the development of plant and animal life and homo sapiens species. The connection of knowledge in architecture and design with research in biology, sociology and the interest of people, companies and government give rise to natural landscapes in urban spaces (**Figures 7–13**) with a meaning of spontaneous social interaction and a dream of an open heritage for the pleasant experience in the community that is essential for human health [12].

Figure 7.
Large urban center square with sustainable landscaping project.

Figure 8.
Sustainable landscaping design in parks within large cities.

Figure 9.
Sustainable landscaping project for urban condominiums.

Figure 10.
Young people practicing physical activity in a park with sustainable landscaping.

Figure 11.
Vertical garden design in modernized urban centers.

The green roof favors the region's biome as it is home to wild animals that can make the neighborhood more pleasant and beautiful (**Figure 14**). The green roof reduces the temperature at the top of a building by up to 5°C, in addition to minimizing the heat islands in cities. It is one of the alternatives to save money, as the captured rainwater can be used to clean environments, gardening and also in the bathroom, in toilets. When treated, filtered and purified, this water can be made potable, stored and sold. The consumption of water captured on green roofs is a benefit for everyone because there is a decrease in the consumption of drinking water [13].

Figure 12.
Sustainable landscaping project for urban residence.

Figure 13.
Sustainable garden for homes.

Figure 14.
Green roof in sustainable landscaping projects.

The construction of green roofs reduces flooding on streets and avenues, as the detention basin on the roofs makes it possible to reduce flooding in large cities, where there are few green areas and a lot of paved area, making it difficult for water to flow in the rainy season [14].

Landscaping is rich in sensory stimuli, as nature has its constant and unpredictable movements. They are the ones that involve our senses and contribute to well-being. The backyard design in the landscaping brings the connection with nature in a mentally and physically restorative space and with the availability of fresh and natural fruits available for naturally healthy eating. A natural refuge that envelops our senses and from which it is possible to observe the surroundings, but from a protected position, far from the hustle and bustle of the city [10].

4. Landscaping architecture

Sustainable architecture means planning including organic and conscious actions in the external space of houses, buildings and buildings, resulting in sustainable landscaping (**Figure 15**). The daily practice of professionals and offices with a focus on sustainability in landscaping projects must have the natural resources to balance the environment in urban areas and in spaces of common use (**Figures 16** and **17**). Sustainable architecture projects beautiful, pleasant, spacious, harmonious scenarios (**Figures 18–22**) to allow a functional routine and total respect for nature and inhabitants [10].

The distribution of plants in an architectural way in areas of high declivity minimizes problems such as floods and landslides in urban centers, being one of the economically viable and sustainable alternatives for heritage preservation in cities [15].

4.1 The oriental garden in sustainable landscaping

The oriental garden is a powerful source of tranquility and serenity in the modern world. Oriental design is perfect for creating interiors and exteriors that stand out for

Figure 15.
Building with innovative architectural design and vertical garden.

Figure 16.
Balcony of building with innovative architectural design and vertical garden.

Figure 17.
Ground floor garden of building with innovative architectural design and vertical garden.

the feeling of peace, balance and contemplation capable of stimulating and transmitting (**Figure 23**). In this type of garden, the choice and arrangement of resources are carried out with care and the elements have their role and function in the composition of the landscape [6].

Oriental gardens are based on balance with the good use of natural elements, excellence in landscaping and in the choice of accessories. Contemplating the oriental gardens is like living and a private oasis dedicated to recharging energies. The assembly of the oriental garden can be carried out by specialized companies. The oriental

Figure 18.
Sustainable landscaping project with innovative architecture.

Figure 19.
Landscaping project in buildings with innovative architecture design.

Figure 20.
Sustainable landscaping project for the interior of large.

Figure 21.
Sustainable landscaping project developed for the Amazon office complex.

garden is the reproduction of some elements of the planet and its natural resources. The stones represent the mountains and islands, the sand can be associated with the passage of time, while the paths and passages refer to the journey of life itself. The garden is inspired by the culture of the East, symbolism is the striking aspect in the design of the oriental garden. The presence of the water element is of fundamental importance in the design of oriental gardens and is associated with calming feelings and mental relaxation and appears in the form of fountains, amphorae, lakes, wells and waterfalls. The lakes characteristic of oriental gardens play an important role in meditation [6].

Bridges carry symbolism and esthetic appeal in the composition of oriental gardens and can be used for better use of space and also for contemplation. Stone laying

Figure 22.
Vertical forest design using modern engineering techniques.

Figure 23.
Natural elements of the oriental garden in sustainable landscaping projects.

is considered an art form in Asia and can be used to highlight the flower bed, plants and water features. Stones also help to direct the focal point in oriental gardens. The selection of decorative stones of varying sizes helps to reproduce the natural environment in the garden and promote relaxation and mental tranquility [6, 16].

Trees are especially important, providing shade and exploration of vertical space in the eastern garden. Japanese maple is one of the species traditionally found in oriental gardens. The choice of plants is based on the color, fragrance and texture of the species. Lilies, Nandina domestica and peonies are some of the popular choices in this type of landscape. Bamboo is the traditional plant of the oriental garden, being used in the construction of gates, flowerbeds, fountains and other accessories [3, 17].

5. Preservation of the physical structure of the soil

The sustainable landscaping reduces damage to residential, commercial and public buildings. Erosion caused by rainfall results in property degradation. Nutrient loss, soil compaction, and reduced biodiversity of soil organisms can severely limit landscaping vitality. Sedimentation caused by erosion increases turbidity levels, which degrades aquatic habitats, and the accumulation of sediments in water channels can reduce flow capacity, causing flooding that floods streets and houses. The design of landscaping architecture (**Figure 24**) optimizes rainwater runoff and makes it sustainable to invest in landscaping projects in areas with problems in the physical structure of the soil [1].

Figure 24.
Preservation of the physical structure of the soil with cover of plants distributed in the design of the square.

6. Conclusions

Sustainable landscaping makes it possible to reuse rainwater for irrigation of plants, the cultivation of native species in cities, soil conservation, energy and water savings, thermal comfort and esthetics for residential, commercial, public and corporate developments.

Cities with sustainable landscaping projects have a decrease in temperature, an increase in relative humidity, an oxygenated atmosphere, noise filtering and more water available for use. Sustainable gardens educate people to care for nature and to live in high natural performance.

Acknowledgements

The authors are grateful to author service Manager IntechOpen.

Conflict of interest

The authors declare no conflict of interest.

Notes/thanks/other declarations

Thanks to all the authors of the articles and books that were used as reference in the writing of this chapter.

Author details

Fernanda Da Conceição Moreira[1*] and Eglerson Duarte[2]

1 Oliveira & Moreira Construction LTDA, Diamantina, Brazil

2 FCIT, Montes Claros, Brazil

*Address all correspondence to: fernandamoreiradtna@gmail.com

References

[1] Hsu CY, Ou SJ. Innovative practice of sustainable landscape architecture education—Parametric-aided design and application. Sustainability. 2022;**14**:4627. DOI: 10.3390/su14084627

[2] King J, Holmes R, Burkholder S, Holzman J, Suedel B. Advancing nature-based solutions by leveraging engineering with nature® strategies and landscape architectural practices in highly collaborative settings. Integrated Environmental Assessment and Management. 2022;**18**:108-114. DOI: 10.1002/ieam.4473

[3] Waldheim C. Landscape as Urbanism: A General Theory. Princeton: Princeton University Press; 2016. DOI: 10.1515/9781400880546

[4] Xi J, Wang X. Development of landscape architecture design students' pro-environmental awareness by project-based learning. Sustainability. 2022;**14**:2164. DOI: 10.3390/su14042164

[5] Kuper R. Policy brief: Alternatives to In-person American Society of Landscape Architects Conferences on landscape architecture. Landscape Journal. 2022;**41**(1):77-93. DOI: 10.3368/lj.41.1.77

[6] Myeong-Jun L. Back to basics in landscape architecture pedagogy through collage and model-making: A qualitative study in Korean basic-level design education. Landscape Research. 2022;**2022**:959-979. DOI: 10.1080/01426397.2022.2089641

[7] Nickayin SS. Paradigm shift of scale in landscape architecture—Towards a planetary observation. Sustainability. 2022;**14**:2949. DOI: 10.3390/su14052949

[8] Ranja H, Julia D. Modern living in a forest—Landscape architecture of Finnish forest suburbs in the 1940s–1960s. Geografiska Annaler: Series B, Human Geography. 2022;**104**(3):250-268. DOI: 10.1080/04353684.2021.1989320

[9] Taufen A, Yangn Y. Sustainable cities and landscapes cultivating infrastructures of health. In: Taufen A, Yangn Y, editors. The Routledge Handbook of Sustainable Cities and Landscapes in The Pacific Rim. 1st ed. Routledge: Taylor Francis Group; 2022. pp. 1-13. DOI: 10.4324/9781003033530-1

[10] Dane C, Mariel CA. Trajectories of practice across time: Moving beyond the histories of landscape architecture. Landscape Research. 2022;**47**(1):25-34. DOI: 10.1080/01426397.2021.1989393

[11] Nikologianni A, Plowman T, Brown BA. Review of embodied carbon in landscape architecture. Practice and Policy. 2022;**8**(22):1-15. DOI: 10.3390/c8020022

[12] Dai Y. Application of regional culture in landscape architecture design under the background of data fusion. Scientific Programming. 2022;**2022**:1-12. DOI: 10.1155/2022/6240313

[13] Gillian L, Sudipto R. Learning and teaching academic standards in landscape architecture. Landscape Research. 2022;**47**(7):936-958. DOI: 10.1080/01426397.2022.2089640

[14] Gramkow MC, Merit MT, Stigsdotter UK. A qualitative study on how Danish landscape architectural firms understand and work with accessibility. Archnet-IJAR. 2022;**16**:536-553. DOI: 10.1108/ARCH-08-2021-0233

[15] Fornal-Pienak B, Mária B. Evaluation of current landscape architecture

approaches in chosen cities in Poland and Slovakia. Acta Horticulturae et Regiotecturae. 2022;**25**(1):28-36. DOI: 10.2478/ahr-2022-0004

[16] Holmes R, Burkholder S, Holzman J, King J, Suedel B. Integrating engineering with nature® strategies and landscape architecture techniques into the Sabine-to-Galveston coastal storm risk management project. Integrated Environmental Assessment and Management. 2022;**18**:63-73. DOI: 10.1002/ieam.4434

[17] Karsten J, Richard S, Elke M, Nilgül K. Teaching landscape architecture: A discipline comes of age. Landscape Research. 2022;**47**(2):167-178. DOI: 10.1080/01426397.2020.1849588